ENTERTAINING MATHEMATICAL PUZZLES

ENTERTAINING MATHEMATICAL PUZZLES

Martin Gardner

Illustrated by Anthony Ravielli

DOVER PUBLICATIONS, INC.
NEW YORK

This Dover edition, first published in 1986, is an unabridged and slightly corrected republication of the work first published by Thomas Y. Crowell Company, New York, in 1961 under the title *Mathematical Puzzles*.

Manufactured in the United States of America
Dover Publications, Inc., 31 East 2nd Street, Mineola, N.Y. 11501

Library of Congress Cataloging-in-Publication Data

Gardner, Martin, 1914–
  Entertaining mathematical puzzles.

  Reprint. Originally published: Mathematical puzzles. New York : Crowell, 1961. With minor corrections.
  Bibliography: p.
  1. Mathematical recreations.   I. Title.
QA95.G29   1986   793.7'4   86-16505
ISBN 0-486-25211-6

FOR JIMMY

Introduction

In selecting material for this collection I have done my best to find puzzles that are unusual and entertaining, that call for only the most elementary knowledge of mathematics, but at the same time provide stimulating glimpses into higher levels of mathematical thinking.

The puzzles (many of which appeared in my column "On the Light Side" that ran in *Science World*) have been grouped into sections, each dealing with a different area of mathematics. Brief comments at the beginning of each section suggest something of the nature and importance of the kind of mathematics one must use in tackling the puzzles of that section. In the answers, I have tried to go into as much detail as space permits in explaining how each problem is solved, and pointing out some of the inviting paths that wind away from the problems into lusher areas of the mathematical jungle.

Perhaps in playing with these puzzles you will discover that mathematics is more delightful than you expected. Perhaps this will make you want to study the subject in earnest, or less hesitant about taking up the study of a science for which a knowledge of advanced mathematics will eventually be required.

Surely no one today can doubt the enormous practical value of mathematics. Without its use as a tool, the discoveries and achievements of modern science would have been impossible. But many people do not realize that mathematicians actually *enjoy* mathematics. Take my word for it, there is as much satisfaction in knocking over an interesting problem with a well-aimed thought as there is in knocking over ten wooden pins with a well-aimed bowling ball.

In one of L. Frank Baum's funniest fantasies, *The Emerald City of Oz*, Dorothy (together with the Wizard and her uncle and aunt) visit the city of Fuddlecumjig in the Quadling section of Oz. Its remarkable inhabitants, the Fuddles, are made of pieces of painted wood cleverly fitted together like three-dimensional jigsaw puzzles. As soon as an outsider approaches they scatter in a heap of disconnected pieces on the floor so that the visitor will have the pleasure of putting them together again. As Dorothy's party leaves the city, Aunt Em remarks:

"Those are certainly strange people, but I really can't see what use they are, at all."

"Why, they amused us for several hours," replies the Wizard. "That is being of use to us, I'm sure."

"I think they're more fun than playing solitaire or mumbletypeg," Uncle Henry adds. "For my part, I'm glad we visited the Fuddles."

I hope that you will resist mightily the temptation to look at the answer before you try seriously to work a problem. And I hope that when you finish with these puzzles you will be glad, like Uncle Henry, to have been befuddled by them.

Martin Gardner

Contents

PART I

ARITHMETIC

PUZZLES

Arithmetic Puzzles

THE NUMBERS THAT are used in counting (1, 2, 3, 4 . . .) are called *integers*. Arithmetic is the study of integers with respect to what are known as the four *fundamental operations of arithmetic:* addition, subtraction, multiplication, and division. (Lewis Carroll's Mock Turtle, you may remember, called them Ambition, Distraction, Uglification, and Derision.) Arithmetic also includes the operations of raising a number to a higher *power* (multiplying it by itself a certain number of times), and of extracting a *root* (finding a number which, when multiplied by itself a certain number of times, will equal a given number).

It goes without saying that you will never be able to learn algebra or any higher branch of mathematics without knowing your arithmetic well. But even if you never learn algebra, you will find that arithmetic is essential to almost every profession you can think of. A waitress has to add the items on a check, a farmer has to calculate the yield of his crops. Even a shoeshine boy must be able to make change correctly, and making change is pure arithmetic. It is as important in daily life as knowing how to tie your shoelaces.

The puzzles in this section and the two that follow call for nothing more than the ability to do simple arithmetic; and to think clearly about what you are doing.

THE COLORED SOCKS

TEN RED SOCKS and ten blue socks are all mixed up in a dresser drawer. The twenty socks are exactly alike except for their color. The room is in pitch darkness and you want two matching socks. What is the smallest number of socks you must take out of the drawer in order to be certain that you have a pair that match?

SOLUTION

Many people, trying to solve this puzzle, say to themselves, "Suppose the first sock that I remove is red. I need another red one to match it, but the next sock might be blue, and the next one, and the next one, and so on until all ten blue socks are taken from the drawer. The *next* sock has to be red, so the answer must be twelve socks."

But something is overlooked in this reasoning. The socks do not have to be a red pair. It is only necessary that they *match*. If the first two fail to match, then the third is sure to match one of the other two, so the correct answer is three socks.

WEIGHTY PROBLEM

IF A BASKETBALL weighs 10½ ounces plus half its own weight, how much does it weigh?

SOLUTION

Before answering this puzzle, it is necessary to know exactly what the words mean. One might, for example, approach it this way: "The basketball weighs 10½ ounces. Half its weight would then be 5¼ ounces. We add these values together to get an answer of 15¾ ounces."

But the problem is to find the weight of the basketball, and if this turns out to be 15¾ ounces, then it cannot also be 10½ ounces as first assumed. There clearly is a contradiction here, so we must have misinterpreted the language of the question.

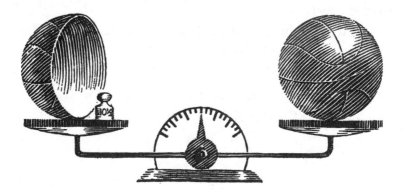

There is only one interpretation that makes sense. The basketball's weight is equal to the sum of two values: 10½ ounces and an unknown value that is half the basketball's weight. This can be pictured on a balance scale as shown in the illustration on the opposite page.

If half a basketball is taken from each side of the scale, the pans will still balance. A 10½-ounce weight will be on one side and half a basketball on the other, so half a basketball must weigh 10½ ounces and the whole basketball must weigh twice this, or 21 ounces.

Actually, without knowing it, we have solved the problem by simple algebra! Instead of pictures, let us represent half a basketball by the letter x. And instead of showing two sides of a scale in balance, let us use the algebraic sign of equality. We can now write the simple equation:

$$10\frac{1}{2} + x = x + x$$

If the same amount is taken from each side of this equation it will still "balance." So we remove x from each side and are left with:

$$10\frac{1}{2} = x$$

You remember that x represented half a basketball. If half a basketball weighs 10½ ounces, then the entire basketball must weigh 21 ounces.

THE SILVER BAR

A SILVER PROSPECTOR was unable to pay his March rent in advance. He owned a bar of pure silver, 31 inches long, so he made the following arrangement with his landlady. He would cut the bar, he said, into smaller pieces. On the first day of March he would give the lady an inch of the bar, and on each succeeding day he would add another inch to her amount of silver. She would keep this silver as security. At the end of the month, when the prospector expected to be able to pay his rent in full, she would return the pieces to him.

March has 31 days, so one way to cut the bar would be to cut it into 31 sections, each an inch long. But since it required considerable labor to cut the bar, the prospector wished to carry out his agreement with the fewest possible number of pieces. For example, he might give the lady an inch on the first day, another inch the second day, then on the third day he could take back the two pieces and give her a solid 3-inch section.

Assuming that portions of the bar are traded back and forth in this fashion, see if you can determine the *smallest* number of pieces into which the prospector needs to cut his silver bar.

The prospector can keep his agreement by cutting his 31-inch silver bar into as few as five sections with lengths of 1, 2, 4, 8, and 16 inches. On the first day he gives the landlady the 1-inch piece, the next day he takes it back and gives her the 2-inch piece, the third day he gives her the 1-inch piece again, the fourth day he takes back both pieces and gives her the 4-inch piece. By giving and trading in this manner, he can add an inch to her amount each day for the full month of 31 days.

The solution to this problem can be expressed very neatly in the *binary system* of arithmetic. This is a method of expressing integers by using only the digits 1 and 0. In recent years it has become an important system because most giant electronic computers operate on a binary basis. Here is how the number 27, for example, would be written if we are using the binary system:

11011

How do we know that this is 27? The way to translate it into our decimal system is as follows. Above the digit on the extreme right of the binary number, we write "1." Above the next digit, moving left, we write "2"; above the third digit from the left, we write "4"; above the next digit, "8"; and above the last digit on the left, "16." (See the illustration.) These values form the series 1, 2, 4, 8, 16, 32 . . . , in which each number is twice the preceding one.

16 8 4 2 1

1 1 0 1 1

The next step is to add together all the values that are above 1's in the binary number. In this case, the values are 1, 2, 8, 16 (4 is not included because it is above a 0). They add up to 27, so the binary number 11011 is the same as 27 in our number system.

Any number from 1 to 31 can be expressed in this way with a binary number of no more than five digits. In exactly the same way, any number of inches of silver from 1 to 31 can be formed with five pieces of silver if the lengths of the five pieces are 1, 2, 4, 8, and 16 inches.

The table here lists the binary numbers for each day in March. You will note that on March 27 the number is 11011. This tells us that the landlady's 27 inches of silver will consist of the 1-inch, 2-inch, 8-inch, and 16-inch sections. Pick a day at random and see how quickly you can learn from the chart exactly which pieces of silver will add to an amount that corresponds with the number of the day.

MARCH	16	8	4	2	1
					BINARY NUMBERS FROM 1 TO 31
1					1
2				1	0
3				1	1
4			1	0	0
5			1	0	1
6			1	1	0
7			1	1	1
8		1	0	0	0
9		1	0	0	1
10		1	0	1	0
11		1	0	1	1
12		1	1	0	0
13		1	1	0	1
14		1	1	1	0
15		1	1	1	1
16	1	0	0	0	0
17	1	0	0	0	1
18	1	0	0	1	0
19	1	0	0	1	1
20	1	0	1	0	0
21	1	0	1	0	1
22	1	0	1	1	0
23	1	0	1	1	1
24	1	1	0	0	0
25	1	1	0	0	1
26	1	1	0	1	0
27	1	1	0	1	1
28	1	1	1	0	0
29	1	1	1	0	1
30	1	1	1	1	0
31	1	1	1	1	1

THE THREE CATS

IF THREE CATS catch three rats in three minutes, how many cats will catch 100 rats in 100 minutes?

SOLUTION

The usual answer to this old riddle is as follows. If it takes three cats three minutes to catch three rats, it must take them one minute to catch one rat. And if it takes them a minute for each rat, then the *same three cats* would catch 100 rats in 100 minutes.

Unfortunately, it is not quite that simple; such an answer presupposes something that is certainly not stated in the problem. It assumes that all three cats concentrate their attention on the same rat until they catch him in one minute, then turn their combined attention toward another rat. But suppose that instead of doing this, each cat chases a different rat and takes three minutes to catch it. In this case, three cats would still catch three rats in three minutes. It would take them six minutes to catch six rats, nine minutes to catch nine rats, and 99 minutes to catch 99 rats.

A curious difficulty now faces us. How long will it take those same three cats to catch the 100th rat? If it still takes them the full three minutes to run him down, then it will take three cats 102 minutes to catch 100 rats.

To catch 100 rats in 100 minutes—assuming this is how the cats go about their rat catching—we will certainly need more than three cats and less than four cats.

Of course it is possible that when the three cats gang up on a single rat they can corner him in less than three minutes, but there is nothing in the statement of the riddle that tells us exactly how to measure the time for this operation. The only correct answer to the problem, therefore, is this: The question is ambiguous and cannot be answered without more information about how those cats catch rats.

MRS. PUFFEM'S CIGARETTES

MRS. PUFFEM, a heavy smoker for many years, finally decided to stop smoking altogether. "I'll finish the twenty-seven cigarettes I have left," she said to herself, "and never smoke another one."

It was Mrs. Puffem's practice to smoke exactly two-thirds of each cigarette. It did not take her long to discover that with the aid of some cellophane tape she could stick three butts together to make a new cigarette. With 27 cigarettes on hand, how many cigarettes can she smoke before she gives up the weed forever?

SOLUTION

After smoking the 27 cigarettes, Mrs. Puffem patched together the butts to make 9 more. These 9 cigarettes left enough butts for 3 more smokes; then with the 3 final butts she made one final cigarette. Total: 40 cigarettes. Mrs. Puffem never smoked again; she failed to recover from the strength of her final puff.

PART II

MONEY

PUZZLES

Money Puzzles

"IF YOU'LL GIVE me your water pistol," says little Tommy to his playmate, "I'll let you have my dump truck." This kind of trading is called "bartering." In primitive societies it is the only way in which things can be "bought" and "sold."

Think about it a moment and you will see what a poor system this is. A man who wants to sell his cow and buy a horse will be unable to do so until he happens to meet another man who wants to sell his horse and buy a cow. It may be years before he finds such a man. And suppose a man wants to trade his cow for a sheep that belongs to one friend and a pig that belongs to another. He can't slice his cow in half and trade each half separately! So you see, in any complicated society where many things are bought and sold, it is necessary to have something called *money;* something that can be split up into any amount one wishes, and which has a value that everyone can agree on.

Almost anything can and has been used in the past for money, but today money consists either of coins made of metal or printed paper money. Few uses of arithmetic are more important than learning how to handle money problems. The next five puzzles will test your ability along these lines, and perhaps teach you a few things you did not fully understand before.

SECOND-HAND SCOOTER

BILL SOLD HIS motor scooter to Tom for $100. After driving it around for a few days Tom discovered it was in such a broken-down condition that he sold it back to Bill for $80.

The next day Bill sold it to Herman for $90. What is Bill's total profit?

SOLUTION

This little puzzle never fails to start arguments. Most people take one of the following three positions:

(1) We don't know what the scooter originally cost, so after the first sale we have no way of knowing whether Bill made a profit or not. However, since he bought it back for $80 and sold it again for $90, he clearly made a $10 profit.

(2) Bill sold his scooter for $100 and bought it back for $80. He now has the same scooter plus $20 that he didn't have before, so his profit is $20. We learn nothing from the next sale because we don't know the scooter's real worth, so Bill's total profit is $20.

(3) After Bill buys back the scooter, his profit is $20 as just explained. He now sells it for $10 more than he just paid for it, making an additional profit of $10. Total profit, therefore, is $30.

Which is correct? The answer is that one is just as good as another! In a series of transactions involving the same object, the "total profit" is the difference between what one first paid for it and the amount one has at the finish. For example, if Bill had paid $100 for the scooter, then he ends up with $110 and we can say that his total profit is $10. But because we don't know the scooter's original cost, we have no way of saying what his final profit is.

Each answer is correct if we are willing to accept some meaning, other than the usual one, of the phrase "total profit." Many problems in life are like this. They are called "verbal problems" or "semantic problems" because they have different answers depending on exactly how one understands the words that are important in the problem. They have no "correct" answers until everyone agrees on the same meaning for the terms.

LOW FINANCE

"I SEEM TO have overdrawn my account," said Mr. Green to the bank president, "though I can't for the life of me understand how it could have happened. You see, I originally had $100 in the bank. Then I made six withdrawals. These withdrawals add up to $100, but according to my records, there was only $99 in the bank to draw from. Let me show you the figures."

Mr. Green handed the bank president a sheet of paper on which he had written:

Withdrawals	Amount left on deposit
$ 50	$50
25	25
10	15
8	7
5	2
2	0
$100	$99

"As you see," said Mr. Green, "I seem to owe the bank a dollar."

The bank president looked over the figures and smiled. "I appreciate your honesty, Mr. Green. But you owe us nothing."

"Then there is a mistake in the figures?"

"No, your figures are correct."

Can you explain where the error lies?

There is no reason whatever why Mr. Green's original deposit of $100 should equal the total of the amounts left after each withdrawal. It is just a coincidence that the total of the right-hand column comes as close as it does to $100.

This is easily seen by making charts to show a different series of withdrawals. Here are two possibilities:

Withdrawals	Amount left on deposit
$99	$1
1	0
$100	$1

Withdrawals	Amount left on deposit
$1	$99
1	98
1	97
97	0
$100	$294

As you see, the total on the left must always be $100, but the total on the right can be made very small or very large. Assuming that withdrawals can never involve a fraction of a cent, try to determine the smallest possible total and the largest possible total that the right-hand column can have.

NO CHANGE

"GIVE ME CHANGE for a dollar, please," said the customer.

"I'm sorry," said Miss Jones, the cashier, after searching through the cash register, "but I can't do it with the coins I have here."

"Can you change a half dollar then?"

Miss Jones shook her head. In fact, she said, she couldn't even make change for a quarter, dime, or nickel!

"Do you have any coins at all?" asked the customer.

"Oh yes," said Miss Jones. "I have $1.15 in coins."

Exactly what coins were in the cash register?

SOLUTION

If Miss Jones couldn't change a dollar, then the cash register could not have in it more than one half dollar. If she couldn't change a half dollar, then the register had no more than one quarter, and no more than four dimes. No change for a dime means no more than one nickel, and no change for a nickel means no more than four pennies. So the cash register could not have contained more than:

1 half dollar	$ .50
1 quarter	.25
4 dimes	.40
1 nickel	.05
4 pennies	.04
	$1.24

A dollar's change can still be made with these coins (for example: a half dollar, quarter, two dimes, and a nickel), but we know that the register cannot have *more* coins than those listed above. They add to $1.24 which is just 9 cents more than $1.15, the amount we are told is in the register.

Now the only way to make 9 cents is with a nickel and four pennies, so those are the coins that must be eliminated. The remaining coins—a half dollar, quarter, and four dimes—will not provide change for a dollar or any smaller coin, and they add to $1.15, so we have found the only answer to the puzzle.

AL'S ALLOWANCE

AL WANTED HIS father to give him an allowance of $1.00 a week, but his father refused to go higher than 50 cents. After they had argued about it for a while, Al (who was pretty smart in arithmetic) said:

"Tell you what, Dad. Suppose we do it this way. Today is the first of April. You give me a penny today. Tomorrow, give me two pennies. The day after tomorrow, give me four pennies. Each day, give me twice as many pennies as you did the day before."

"For how long?" asked Dad, looking wary.

"Just for the month of April," said Al. "Then I won't ask you for any more money for the rest of my life."

"Okay," Dad said quickly. "It's a deal!"

Which of the following figures do you think comes the closest to the amount of money that Dad will have to pay Al during the month of April?

$1
$10
$100
$1,000
$10,000
$100,000
$1,000,000
$10,000,000

If you keep doubling a penny, the amount starts to grow slowly at first, then faster and faster until soon it gallops along with enormous leaps. It is hard to believe, but if poor Dad keeps his agreement he will have to pay Al more than ten million dollars!

On the first day Dad pays Al a penny. The next day, 2 pennies, making a total of 3. The third day he gives his son 4 pennies, raising the total to 7. Let's make a chart to show this for the first week:

Day of month	Pennies for that day	Total pennies
1	1	1
2	2	3
3	4	7
4	8	15
5	16	31
6	32	63
7	64	127

If this chart is continued it will show that Dad's final payment, on April 30, is $5,368,709.12, or well over five million dollars. This, however, is only Dad's *last* payment. We still need to know how much he pays altogether, and to get this we must add all thirty of his payments. This can be done quickly by using the following short cut.

Note that each number in the right-hand column

of the chart is just one less than twice the corresponding number in the center column. So all one has to do is double Dad's last payment to get \$10,737,418.24, then subtract 1 penny to get \$10,737,418.23. This is the total amount Dad will have to fork over if he keeps his agreement.

PICK YOUR PAY

Suppose you take a new job and the boss offers you a choice between:

(A) $4,000 for your first year of work, and a raise of $800 for each year thereafter;

(B) $2,000 for your first six months of work, and a raise of $200 every six months thereafter.

Which offer would you take and why?

SOLUTION

Surprisingly, the second offer is much better than the first one. If you accept it, you will earn exactly $200 *more* each year than you would on the other basis of payment! The following chart shows your total earnings, on the basis of both offers, for each of the first six years.

Years	Offer A	Offer B
1	$4,000	$4,200
2	4,800	5,000
3	5,600	5,800
4	6,400	6,600
5	7,200	7,400
6	8,000	8,200

PART III

SPEED

PUZZLES

Speed Puzzles

WE LIVE IN a world in which everything is always changing, though in ten thousand different ways and at all sorts of different speeds. The sky may darken in a few hours, a banana darkens in a few days. The colors on wallpaper fade so slowly that it may be years before we notice the change. Some changes are extremely irregular, like the way you change positions when you sleep. Other changes, such as the waxing and waning of the moon, or the vibration of an atom in a molecule, are more regular than clockwork.

The branch of mathematics that is most concerned with change is called the *calculus*. It is impossible to be a physicist today without knowing calculus; but, before you can understand it, you must first know a great deal about the mathematics of simple and regular types of change that can be handled by ordinary arithmetic. The most common example of such a change is the change of position that we call *constant speed*. It is expressed as a ratio between distance and time:

$$\text{Speed} = \frac{\text{Distance}}{\text{Time}}$$

With this basic formula in mind, and some hard clear thinking, perhaps you will be able to master the four unusual speed problems that follow.

THE BICYCLES AND THE FLY

Two boys on bicycles, 20 miles apart, began racing directly toward each other. The instant they started, a fly on the handle bar of one bicycle started flying straight toward the other cyclist. As soon as it reached the other handle bar it turned and started back. The fly flew back and forth in this way, from handle bar to handle bar, until the two bicycles met.

If each bicycle had a constant speed of 10 miles an hour, and the fly flew at a constant speed of 15 miles an hour, how far did the fly fly?

SOLUTION

Each bicycle travels at 10 miles an hour, so they will meet at the center of the 20-mile distance in exactly one hour. The fly travels at 15 miles an hour, so at the end of the hour it will have gone 15 miles.

Many people try to solve this the hard way. They calculate the length of the fly's first path between handle bars, then the length of his path back, and so on for shorter and shorter paths. But this involves what is called the *summing of an infinite series,* and it is very complicated, advanced mathematics.

It is said that the Hungarian mathematician, John

von Neumann, perhaps the greatest mathematician in the world when he died in 1957, was once asked this problem at a cocktail party. He thought for a moment, then gave the correct answer. The person who asked the question looked a bit crestfallen. He explained that most mathematicians overlook the simple way it can be solved and try to solve it by the lengthy process of summing an infinite series.

Von Neumann looked surprised. "But that's how I solved it," he explained.

THE FLOATING HAT

A FISHERMAN, wearing a large straw hat, was fishing from a rowboat in a river that flowed at a speed of three miles an hour. His boat drifted down the river at the same rate.

"I think I'll row upstream a few miles," he said to himself. "The fish don't seem to be biting here."

Just as he started to row, the wind blew off his hat and it fell into the water beside the boat. But the fisherman did not notice his hat was gone until he had rowed upstream five miles from his hat. Then he realized what must have happened, so he immediately started rowing back downstream again until he came to his floating hat.

In still water, the fisherman's rowing speed is always five miles an hour. When he rowed upstream and back, he rowed at this same constant speed, but of course this would not be his speed relative to the *shore* of the river. For instance, when he rowed upstream at five miles an hour, the river would be carrying him downstream at three miles an hour, so he would be passing objects on the shore at only two miles an hour. And when he rowed downstream, his rowing speed and the speed of the river would combine to make his speed eight miles an hour with respect to the shore.

If the fisherman lost his hat at two o'clock in the afternoon, what time was it when he recovered it?

Because the speed of the river has the same effect on both boat and hat, it can be ignored completely in solving this puzzle. Instead of the water moving and the shore remaining fixed, imagine the water as perfectly still and the shore moving. As far as boat and hat are concerned, this situation is exactly the same as before. Since the man rows five miles away from the hat, then five miles back to the hat, he has rowed a total distance of ten miles *with respect to the water*. His rowing speed with respect to the water is five miles an hour, so it must have taken him two hours to go the ten miles. He would recover his hat, therefore, at four o'clock.

The situation here is comparable to that of calculating speeds and distances on the surface of the earth. The earth is spinning through space, but because this motion has the same effect on all objects on its surface, it can be ignored completely in most speed and distance problems.

ROUND TRIP

WHEN A TRIP is made by car, the car will of course travel different speeds at different times. If the total distance is divided by the total driving time, the result is called the *average speed* for that trip.

Mr. Smith planned to drive from Chicago to Detroit, then back again. He wanted to average 60 miles an hour for the entire round trip. After arriving in Detroit he found that his average speed for the trip was only 30 miles an hour.

What must Smith's average speed be on the return trip in order to raise his average for the round trip to 60 miles an hour?

SOLUTION

It is not necessary to know the distance between Chicago and Detroit in order to solve this perplexing little puzzle. When Smith arrived in Detroit he had gone a certain distance and it had taken him a certain length of time. If he wishes to double his average speed, it is necessary for him to go *twice* that distance in the *same* length of time. Clearly, in order to do this he must return to Chicago in *no time at all!* This is impossible, so there is no way Smith can raise his average speed to 60 miles an hour. No matter how fast he makes the return trip he is sure to fall short of a 60-mile-per-hour average.

This will be easier to understand if we pick a certain distance for Smith to travel; say 30 miles and back. Since his average speed is 30 miles an hour, he will complete the first half of his trip in one hour. He wants to make the round trip with an average speed of 60 miles an hour, which means that he must complete the entire 60-mile trip in one hour. But he has already used up his hour. No matter how fast he returns, his total time will be more than one hour, therefore he will have traveled 60 miles in more than an hour, making his average speed less than 60 miles an hour.

AIRPLANE PARADOX

AN AIRPLANE FLIES from city A to city B, then back to A again. When there is no wind, its average ground speed (speed relative to the ground) for the entire trip is 100 miles per hour. Assume that a steady wind is blowing in a straight direction from city A toward city B. How will this wind affect the plane's average ground speed for the round trip, assuming that it flies at all times with the same motor speed as before?

Mr. White argues: "It won't affect the average speed at all. The wind will speed up the plane on its flight from A to B, but on the return trip it will slow down the plane by the same amount."

"That sounds reasonable," agrees Mr. Brown, "but suppose the wind is 100 miles an hour. The plane will go from A to B at 200 miles an hour, but its return speed will be zero! The plane won't be able to get back at all!"

Can you explain this seeming paradox?

SOLUTION

Mr. White is right in saying that the wind increases the plane's speed in one direction by the same amount that it decreases the speed in the other direction. But he is wrong when he says the wind will not affect the plane's average ground speed for the entire round trip.

What Mr. White failed to consider was the length of time that the plane flies at each of the two speeds. The return trip against the wind will take much longer than the trip with the wind. As a result, more time is spent in flying at the reduced ground speed, and so the average ground speed for both trips will be *less* than if there were no wind. The stronger the wind, the greater this reduction will be. When the speed of the wind equals or exceeds the plane's speed, then average ground speed for the round trip becomes zero because the plane is unable to return.

PART IV

PLANE

GEOMETRY

PUZZLES

Plane Geometry Puzzles

IF WE WANTED to be very up to date and technical we could define geometry by quoting this definition: "The study of invariant properties of given elements under specified groups of transformations." But to understand that, you would have to know what all the words mean and some of them are not easy to explain. So we will take a less technical approach and simply say that geometry studies the sizes and shapes of things.

Plane geometry is the most elementary branch of geometry. It deals with the mathematical properties of flat figures, such as lines, angles, triangles, squares, and circles, that can be drawn on a sheet of paper with a ruler and a compass. It had its beginnings in ancient Egypt, but it was the Greeks who first developed it into a science. They were interested in plane geometry not only because it was useful in surveying and carpentry and architecture, but also because of its great beauty. No man could call himself truly educated, the Greeks believed, who did not understand some geometry.

The next four puzzles do not require any special knowledge of plane geometry, but they will test your ability in the kind of pictorial thinking that is so useful in solving geometrical problems.

CORNER TO CORNER

MANY TIMES A geometrical problem is enormously difficult if it is approached the wrong way. Tackle it another way and it is absurdly simple. This problem is a classic example.

Given the dimensions (in inches) shown in the illustration, how quickly can you compute the length of the rectangle's diagonal that runs from corner A to corner B?

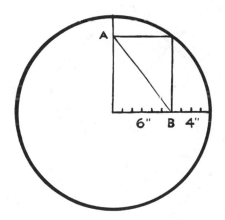

SOLUTION

Draw the other diagonal of the rectangle and you will see at once that it is the radius of the circle. The diagonals of a rectangle are always equal, therefore the diagonal from corner A to corner B is equal to the circle's radius, which is 10 inches!

THE HINDU AND THE CAT

How MANY DIFFERENT squares can you count in the picture of the turbaned Hindu boy?

How many different triangles can you count in the picture of the cat?

Look carefully. The problems are not as easy as you might think!

SOLUTION

In working on problems of this sort it is always best to count the figures in some systematic way. In the picture of the Hindu boy, let's take the squares in order of size:

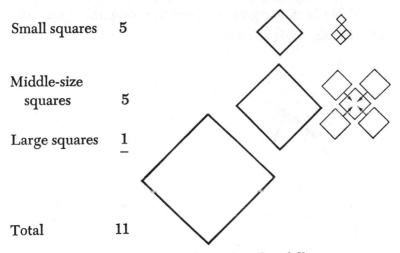

Small squares	5
Middle-size squares	5
Large squares	1
Total	11

The triangles in the cat can be counted as follows:

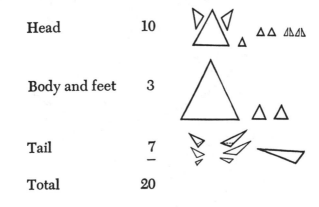

Head	10
Body and feet	3
Tail	7
Total	20

CUTTING THE PIE

WITH ONE STRAIGHT cut you can slice a pie into two pieces. A second cut that crosses the first one will produce four pieces, and a third cut (see the illustration) can produce as many as seven pieces.

What is the largest number of pieces that you can get with six straight cuts?

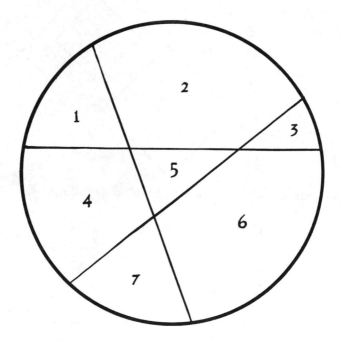

Instead of solving this puzzle by trial and error, a better way is to discover a rule that will give the largest number of pieces that can be obtained with any number of cuts.

The uncut pie is one piece, so when cut No. 1 is made, one more piece of pie is added to make two pieces in all.

Cut No. 2 adds 2 more pieces, making 4 in all.

Cut No. 3 adds 3 more pieces, making 7 in all.

It looks as if each cut adds a number of pieces that is always equal to the number of the cut. This is true, and it is not hard to see why. Consider, for example, the third cut. It crosses two previous lines. Those two lines will divide the third line into three sections. Each of those three sections cuts a piece of pie into two parts, so each section will add one extra piece and the three sections naturally add three pieces.

The same is true of the fourth line. It can be drawn so it crosses the other three lines. These three lines will divide the fourth line into four sections. Each section adds an extra piece so the four sections will add four more pieces. And the same is true of the fifth line, sixth line, and so on for as many lines as we care to add. This type of reasoning, from particular cases to an infinite number of cases, is known as *mathematical induction*.

Bearing the rule in mind, it is now a simple matter to make a chart showing the largest number of pieces that each cut will produce:

Number of cuts	Number of pieces
0	1
1	2
2	4
3	7
4	11
5	16
6	22

How many pieces can you make with seven cuts? We simply add 7 to 22 and we know the answer is 29. The illustration shows how six cuts can be made to produce 22 pieces, the answer to the original problem.

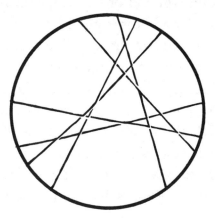

WHERE DOES THE SQUARE GO?

PAUL CURRY, an amateur magician in New York City, was the first to discover that a square can be cut into a few pieces and the pieces rearranged to make a square of the same size that has a hole in it!

There are many versions of the Curry paradox, but the one shown in Figures 1 and 2 is one of the simplest. Paste a sheet of graph paper on a piece of cardboard. Draw the square shown in Figure 1, then cut along the lines to make five pieces. When you rearrange these same five pieces, in the manner shown in Figure 2, a hole will appear in the center of the square!

The square in Figure 1 is made up of 49 smaller squares. The square in Figure 2 has only 48 small squares. Which small square has vanished and where did it go?

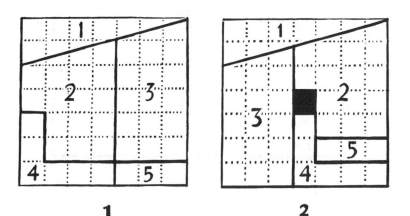

1   2

When the two largest pieces are switched, each small square that is cut by the diagonal line becomes a trifle higher than it is wide. This means that the large square is no longer a perfect square. It has increased in height by an area that is exactly equal to the area of the hole.

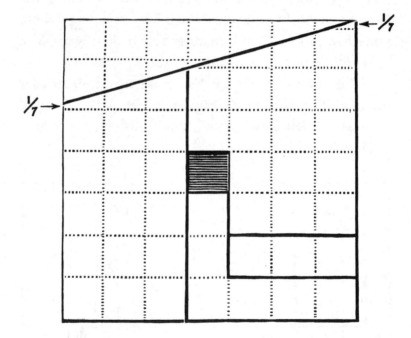

PART V

SOLID
GEOMETRY
PUZZLES

Solid Geometry Puzzles

WHEN WE TURN from plane to solid geometry, we turn from the flat two-dimensional world of a sheet of paper or a TV screen to the rich three-dimensional world of everyday life. Our bodies are three-dimensional. Our houses are three-dimensional. We live on a 3-D solid that is a sphere slightly flattened at the poles and a trifle pear shaped. Solid geometry studies the shapes and sizes of all three-dimensional things.

You may have noticed that many familiar two-dimensional figures have their close cousins in three dimensions. On the plane a compass traces a *circle*. In the air, if we keep the point of the compass in a fixed position and allow the pencil end to swing in all directions (or if we rotate a circle), it will range over the surface of a *sphere*. When a Beatnik wants to describe someone as "squarer" than a "square," he calls him by the name of the square's three-dimensional counterpart, the *cube*. The equilateral triangle also has its 3-D counterpart, the *tetrahedron*. It is a pyramid with four faces, each of which is an equilateral triangle.

The ability to think in three dimensions, tested by the four puzzles in this section, is of great importance in almost every science.

UNDER THE BAND

IMAGINE THAT YOU are on a perfectly smooth sphere as big as the sun. A steel band is stretched tightly around the equator.

One yard of steel is added to this band so that it is raised off the surface of the sphere by the same distance all the way around. Will this lift the band high enough so that you can:

> (1) Slip a playing card under it?
> (2) Slip your hand under it?
> (3) Slip a baseball under it?

SOLUTION

It seems surprising, but that steel band, after a yard has been added to it, will be raised almost six inches all the way around! This is certainly high enough for a baseball to pass underneath.

Actually, the height the band is raised is the same regardless of how large the sphere may be. It is easy to see why. When the band is tight around the sphere, it is the circumference of a circle with a radius that is the same as the radius of the sphere. We know from plane geometry that the circumference of a circle is equal to its diameter (which is twice its radius) times pi. Pi is

3.14+, a number that is a little more than 3. Therefore, if we increase the circumference of *any* circle by one yard, we must increase the diameter by a trifle less than one-third of a yard, or almost a foot. This means, of course, that the radius will increase by almost six inches.

As the illustration makes clear, this increase in radius is the height that the band will be raised from the sphere's surface. It will be exactly the same, 5.7+ inches, regardless of whether the sphere be as large as the sun or as small as an orange!

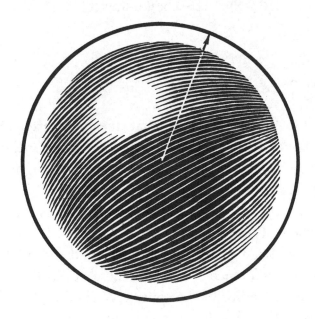

THE THIRD LINE

A STRAIGHT LINE is called *self-congruent* because any portion of the line can be exactly fitted to any other portion of the same length. The same is true of the circumference of a circle. Any part of the circumference is exactly like any other part of the same length. An oval line is not self-congruent because parts of it have different curvature. A portion of an oval taken from the side would not fit the more sharply curved portion at one of the ends.

There is a third type of line that is self-congruent like the straight line and circle. Can you think of what sort of line it is?

SOLUTION

Because this problem is in a section on solid geometry, perhaps you have guessed that the third type of self-congruent line is one that cannot be drawn on a plane. It is called the *circular helix*—a line that spirals through space like a corkscrew or the lines on a barber's pole. If you study the illustration, you will see that any portion of this helix will fit any other portion.

There are other types of helices, but only the circular helix is self-congruent. The circular helix is one that spirals with a constant angle around a cylinder that has a circular cross section. Other helices are those that spiral around cylinders with noncircular cross sections, and around cones. A cone-shaped bedspring is a familiar example of a conical helix. Helices have many interesting properties, and they are often encountered in physics, astronomy, chemistry, biology, and other sciences.

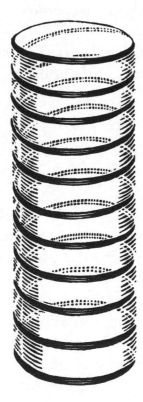

THE PAINTED CUBES

IMAGINE THAT YOU have a can of red paint, a can of blue paint, and a large supply of wooden cubes, all the same size. You decide to paint the cubes by making each face either solid red or solid blue. For example, you might paint one cube all red. The next cube you may decide to give three red faces and three blue faces. Perhaps the third cube can also be given three red and three blue faces, but painted in such a way that it doesn't match the second cube.

How many cubes can you paint in this manner that will be different from each other? Two cubes are considered alike if one can be turned so that all its sides match the corresponding sides of the other cube.

SOLUTION

You can paint:

   1 cube that is all red,
   1 cube that is all blue,
   1 cube with 5 red faces, 1 blue,
   1 cube with 5 blue faces, 1 red,
   2 cubes with 4 red faces, 2 blue.
   2 cubes with 4 blue faces, 2 red,
   2 cubes with 3 red faces, 3 blue.

This makes a total of *ten* different cubes.

THE SPOTTED BASKETBALL

WHAT IS THE largest number of spots that can be painted on a basketball in such a way that every spot is exactly the same distance from every other spot?

"Distance" here means the distance measured on the surface of the sphere. A good way to work on this puzzle is to mark spots on a ball and measure the distances between them with a piece of string.

SOLUTION

No more than four spots can be painted on a sphere so that each spot is the same distance from every other spot. The illustration shows how the spots are placed. It is interesting to note that if we draw straight lines inside the sphere, connecting the centers of the four spots, these lines will mark the edges of a tetrahedron.

PART VI

GAME

PUZZLES

Game Puzzles

Dɪᴅ ʏᴏᴜ ᴇᴠᴇʀ stop to think that a great many games are really mathematical puzzles? Ticktacktoe, for example, is pure mathematics. It is such a simple game that it is not hard to analyze it completely and become a player who never makes a mistake. In modern *game theory*, one of the newest branches of mathematics, such a player is said to play *rationally*. When two ticktacktoe players both play rationally, the game is always a draw.

Chess and checkers are two other familiar examples of mathematical games, but there are so many different ways to make moves that no one has yet completely analyzed either game. If two chess or checker players play rationally, will the game always be a draw or does the first or second player have a sure way to win? Nobody knows. If they did, chess and checkers would be much less interesting games!

The four puzzles in this section are four novel games that are easy to analyze and cannot end in draws. Try playing them with a friend and see how quickly you can discover how the first or second player can always win if he plays correctly.

THE CIRCLE OF PENNIES

To PLAY THIS game, take any number of counters (they can be pennies, checkers, pebbles, or bits of paper) and arrange them in a circle. The illustration shows the start of a game with ten pennies. Players take turns removing one or two counters, but if two are taken they must be *next to each other,* with no counters or open spaces between them. The person who takes the last counter is the winner.

If both sides play rationally, who is sure to win and what strategy should he use?

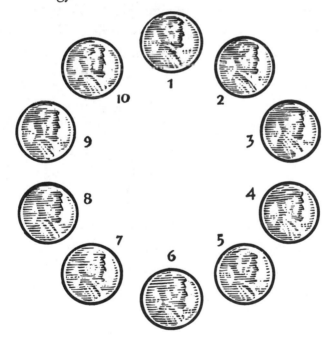

The second player, if he uses the following two-part strategy, can always win this game:

1. After the first player has removed one or two counters, there will be a single gap somewhere in the circle. The second player now takes either one or two counters from the opposite side of the circle so that the counters are left divided into two equal groups.

2. From now on, whatever the first player takes from one group, the second player takes the corresponding counter or counters from the other group.

This strategy will become clear if you play over the following sample game. The numbers refer to the numbers given to each of the coins in the illustration on page 55.

First player	Second player
8	3
1,2	5,4
7	9
6	10 (wins)

Try using this strategy on your friends and you'll soon see why it cannot fail to win for the second player, regardless of how many counters are used in forming the circle.

FOX AND GOOSE

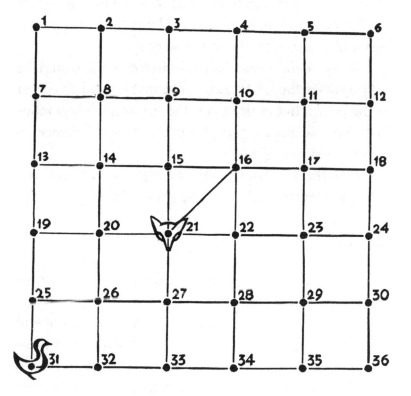

THIS AMUSING GAME is played on the above board.

Place a penny on the picture of the fox and a dime on the picture of the goose.

One player moves the fox, the other moves the goose. A "move" consists of sliding the coin from one dot to an adjacent dot, along a black line. The fox tries to capture the goose by moving onto the spot occupied by the goose. The goose tries to prevent this. If the fox captures the goose in ten moves or less (that is, ten of the fox's moves), then he wins. If he fails to capture the goose in ten moves, the goose wins.

Now, if the goose had the first move it would be very easy for the fox to trap her in that lower left corner of the board. But in this game the fox must always move first. This seems to give the goose a good chance to escape being caught.

Can the fox always capture the goose in ten moves, if he plays correctly, or can the goose always get away?

SOLUTION

The fox can always capture the goose in less than ten moves. This is how it's done:

His first three moves must take him around one of the two triangles in the center of the board. After completing this circuit, it is then a simple matter for him to trap the goose in a corner square before his ten moves are up.

The following game is typical:

Move	Fox	Goose	Move	Fox	Goose
(1)	16	32	(5)	28	32
(2)	22	33	(6)	27	31
(3)	21	27	(7)	26	25
(4)	22	33	(8)	25 (wins)	

BRIDG-IT

THIS UNUSUAL GAME was invented by David Gale, a professor of mathematics at Brown University, and has been marketed under the trade name of *Bridg-It*. It can be played on boards of various sizes. The version explained here is easy to play on paper, with pencils of two different colors. It's more fun than ticktacktoe!

Suppose that the pencils you use are red and black. With the black pencil, make a rectangle of 12 dots as shown in Figure 1. With the red pencil, add 12 more dots as shown in Figure 2. (In these pictures the red dots are shown as shaded circles.) Figure 2 is the board on which the game is played.

One player holds the black pencil, his opponent

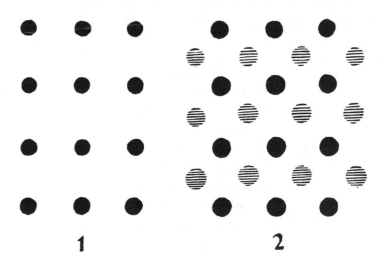

1   2

59

holds the red. The first player draws a horizontal or vertical line that connects two adjacent dots of his own color. Then the other player does the same thing, connecting two adjacent dots of *his* color. They take turns doing this. Black tries to form a continuous path of lines from the top row of black dots to the bottom row. The path does not have to be straight; it can twist in any way so long as it connects opposite sides of the board. Red tries to form a similar path from the left column of red dots to the column of red dots at the right end. Each of course also uses his lines for blocking the other player's path.

The first player to complete his path is the winner. Figure 3 shows the finish of a typical game. Red (whose lines are shown as dotted) has won.

The game cannot end in a draw. Who is sure to win if he plays rationally, the first player or the second player?

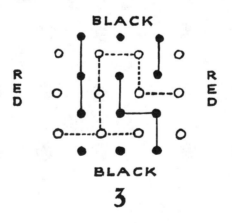

3

A number of opening moves will always win for the first player. One of them is to connect the two dots that are closest to the center of the board. There are too many different lines of play to discuss all of them here, but this move and careful playing thereafter will win the game.

There is an interesting way to prove that the first player, no matter how large the board, can always win if he plays correctly.

It goes like this:

(1) Assume, just for fun, that the *second* player has a sure strategy for winning.

(2) The first player draws his first line anywhere. Then after the second player has drawn a line, the first player pretends that he is the second player, and plays the winning strategy.

(3) The line that the first player made on his first move cannot interfere with his winning strategy. If this line is not part of the strategy, then it doesn't matter. If it is part of the strategy, then when the time comes to draw it, he simply draws a line somewhere else.

(4) Therefore the first player can always win.

(5) But this contradicts our first assumption that the second player could win. Therefore the assumption was wrong.

(6) The game cannot end in a draw, so if there

is no winning strategy for the second player, there must be one for the first player!

This proof, which applies to games other than Bridg-It, is a famous proof in game theory because it shows that there is a winning strategy for the first player, on any size board, but it doesn't explain what that strategy is. The proof is not easy to understand when explained as briefly as it is here, but if you think it through carefully, it should eventually become clear. Mathematicians call it an *existence proof* because it proves that something exists without telling how to go about finding it.

In this case, the type of reasoning used in the proof is known as *reductio ad absurdum*, which is Latin for "reduction to absurdity." You show that one of two things must be true, you assume one to be true, it leads to a logical absurdity, therefore the *other* thing must be true. Here the proof runs as follows: (1) one of two players must win, (2) it is assumed that the second player can always win, (3) this leads to a logical contradiction, (4) therefore, the first player can always win.

This is a powerful form of proof that is often used by mathematicians.

NIM

ARRANGE NINE PENNIES in three rows as shown. Players take turns removing one or more pennies provided they all come from the same row. For example, a player could take one penny from the top row, or *all* the pennies in the bottom row. The person who is forced to take the last penny is the *loser*.

If the first player makes a correct first move, and continues to play rationally, he can always win. If he fails to make this move, his opponent, by playing rationally, can always win.

Can you discover the winning first move?

The only way that the first player can be sure of winning is by taking three pennies from the bottom row on his first move.

Any play that leaves one of the following patterns is sure to win:

1. One coin in each of three rows.
2. Two coins in each of two rows.
3. Three coins in each of two rows.
4. One coin in one row, two in another, three in a third.

If you keep these four winning patterns in mind, you should be able to defeat an inexperienced player every time that you have the first move, as well as every time that he goes first and fails to make the correct opening move.

Nim can be played with any number of counters in any number of rows. The game has been completely analyzed by using the binary system of arithmetic.

It is believed to be Chinese in origin, but the name "Nim" was given to it in 1901 by Charles Leonard Bouton, a professor of mathematics at Harvard University, who was the first to make a complete analysis of the game. "Nim" is an obsolete English word meaning "to steal or take away."

PART VII

PROBABILITY
PUZZLES

Probability Puzzles

EVERYTHING WE DO, everything that happens around us, obeys the laws of probability. We can no more escape them than we can escape gravity. The phone rings. We answer it because we think someone is calling our number, but there is always a chance that the caller dialed the wrong number by mistake. We turn on a faucet because we believe it is probable that water will come out of it, but maybe it won't. "Probability," a philosopher once said, "is the very guide of life." We are all gamblers who go through life making countless bets on the outcome of countless actions.

Probability theory is that branch of mathematics that tells us how to estimate degrees of probability. If an event is certain to happen, it is given a probability of 1. If it is certain not to happen, it has a probability of 0. All other probabilities that lie between 1 and 0 are expressed as fractions. If an event is just as likely to happen as not, we say the probability is ½. Every field of science is concerned with estimating probability. A physicist calculates the probable path of a particle. A geneticist calculates the chances that a couple will have blue-eyed children. Insurance companies, businessmen, stockbrokers, sociologists, politicians, military experts —all have to be skilled in calculating the probability of the events with which they are concerned.

THE THREE PENNIES

JOE: "I'm going to toss three pennies in the air. If they all fall heads, I'll give you a dime. If they all fall tails, I'll give you a dime. But if they fall any other way, you have to give me a nickel."

Jim: "Let me think about this a minute. At least two coins will *have* to be alike because if two don't match, the third will have to match one of the other two. [See the problem of the colored socks in the first section of this book.] And if two are alike, then the third penny will either match the other two or not match them. The chances are even that the third penny will or won't match. Therefore the chances must be even that the three pennies will be all alike or not all alike. But Joe is betting a dime against my nickel that they won't be all alike, so the bet should be in my favor. Okay, Joe, I'll take the bet!"

Was it wise for Jim to accept the bet?

SOLUTION

It was not very wise of Jim to take that bet. His reasoning about it is completely wrong.

To find the chances that the three coins will fall alike or not alike, we must first list all the possible ways that three coins can fall. There are eight such ways, shown in the illustration.

Each way is just as likely to occur as any other way. Note that only two of them show all the coins alike. This means that the chances of all three coins being alike is two out of eight, or ⅛, a fraction that can be simplified to ¼.

There are six ways that the coins can fall without being all alike. Therefore the chance that this will happen is ⅝ or ¾.

In other words, Joe would expect, in the long run, to win three times out of every four tosses. For these wins, Jim would pay him fifteen cents. For the one time that Jim would win, he would pay Jim a dime. This gives Joe a profit of five cents every four tosses—a tidy profit if they kept repeating the bet.

THE TENTH ROLL

AN ORDINARY DIE (such as used in gambling) has six sides, so the probability that any one side will come up is one out of six, or ⅙. Suppose you roll a certain die nine times. Each time the 1-spot turns up.

What is the probability that the 1-spot will turn up again on the next roll? Is it better than ⅙, less than ⅙, or is it still ⅙?

If we know positively that it is a fair die, then no matter how many times it is rolled, or what turns up, the probability on the next roll will still be ⅙ for each of the six faces. A die has no way of remembering what it rolled before!

This is hard for many people to believe. All sorts of foolish systems for playing roulette and other games of chance are based on the superstition that the more often something happens by chance, the less likely it is to happen again. Soldiers in the First World War thought that if they hid in fresh shell holes they were safer than in old ones because, they reasoned, it was unlikely that a shell would explode twice within a short time in exactly the same spot! A mother with five children, all girls, thinks the chances are better than ½ that her next child will be a boy. Both of these beliefs are unfounded.

Now for the other side of the question. In rolling an actual die, it is difficult to be sure that it is not a loaded one, or perhaps controlled by hidden magnets. So if we get an ace on the first nine rolls, we have strong reason for suspecting that the die is what statisticians call a *biased* one. Therefore the probability is *better* than ⅙ that we will get another ace on the tenth roll!

ODDS ON KINGS

Six playing cards are lying face down on the table. You have been told that two and only two of them are kings, but you do not know the positions of the kings.

You pick two cards at random and turn them face up.

Which is the most likely:

(1) There will be at least one king among the two cards?

(2) There will be no king among the two cards?

To solve this problem, let's number the six cards from 1 to 6, and assume that cards 5 and 6 are the two kings.

We now make a list of all the different combinations of two cards that can be picked from the six. There are 15 such combinations:

$$
\begin{array}{lllll}
1\text{-}2 & 2\text{-}3 & 3\text{-}4 & 4\text{-}5 & 5\text{-}6 \\
1\text{-}3 & 2\text{-}4 & 3\text{-}5 & 4\text{-}6 & \\
1\text{-}4 & 2\text{-}5 & 3\text{-}6 & & \\
1\text{-}5 & 2\text{-}6 & & & \\
1\text{-}6 & & & &
\end{array}
$$

Notice that the kings (cards 5 and 6) appear in nine out of the 15 pairs. Since one pair is just as likely as another, this means that, in the long run, you will turn up at least one king in nine out of every 15 tries. In other words, the chance of getting a king is $9/15$, a fraction that simplifies to $3/5$. This of course is better than $1/2$, so the answer to the problem is that you are more likely to get at least one king than no king at all.

What are your chances of finding *both* kings when you turn over two cards? Only one combination among the 15 contains both kings, so the answer is $1/15$.

BOYS vs. GIRLS

George Gamow and Marvin Stern, in their stimulating little book *Puzzle-Math,* tell about a sultan who considered increasing the number of women in his country, as compared to the number of men, so that the men could have larger harems. To accomplish this, he proposed the following law: *as soon as a mother gave birth to her first son, she would be forbidden to have any more children.*

In this way, the sultan argued, some families would have several girls and only one boy, but no family could have more than one boy. It should not be long until the females would greatly outnumber the males.

Do you think the sultan's law would work?

No, the sultan's law would not work.

In obedience to the laws of chance, the first children born to all the women would tend to divide evenly between boys and girls. The mothers of the boys would have no more children. The mothers of the girls would then have their second round of children, and again half would be boys and half girls. Once more the mothers of the boys would drop out, leaving the other mothers to have a third round of children. In each round, the number of girls would tend to equal the number of boys, so the ratio of boys to girls would never change.

"You see," Gamow and Stern write in their answer to the sultan's problem, "that the ratio is maintained. Since in any round of births the ratio of boys to girls is one to one, it follows that when you sum the results of all the rounds, the ratio remains one to one throughout."

Of course while this was going on, girl children would grow up and become new mothers, but the same argument applies to them also.

PART VIII

TOPOLOGY
PUZZLES

Topology Puzzles

TOPOLOGY IS ONE of the youngest and rowdiest branches of modern geometry. Some of its curious figures—one-sided surfaces, closed bottles with no "insides," inner tubes that turn inside out—are so weird that they seem to have been invented by writers of science fiction instead of sober-minded mathematicians.

What is topology? It is the study of properties that remain unchanged regardless of how we twist, stretch, or compress a figure. To a topologist a triangle is the same as a circle because if we imagine the triangle to be made of string, we can easily pull the string into the shape of a circle. Suppose that we have a doughnut (a topologist calls it a *torus*) made of a plastic substance that can be molded in any way we please, but it does not stick to itself and it is impossible to break. You might think that no original features of the doughnut would survive if we pulled, bent, and deformed it enough. But there are many that do survive. For example, it would always have a hole. Such unchangeable properties are its topological properties. They have nothing to do with size, or shape in the sense in which shape is usually understood. They are the deepest of all geometrical properties.

Many puzzles are topological in nature. The following are four of the best.

THE FIVE BRICKS

This is one of the oldest and most famous of all topological puzzles. Your grandfather probably worked on it in school when he was supposed to be studying his history books. Yet not one person in a thousand knows for sure whether it can or can not be done.

The problem is this. Can you draw the diagram in Figure 1 with three strokes of the pencil? You are not permitted to go over any line twice. It's easy to draw all of the figure except for one little segment (a few such attempts are shown in Figure 2), but can the *entire* figure be drawn with three strokes? If not, why not?

The puzzle is topological because the actual sizes and shapes of the bricks do not matter. For example, if we distort the figure as shown in Figure 3, the problem remains exactly the same. Any solution for Figure 1 would be a solution for Figure 3 and vice versa.

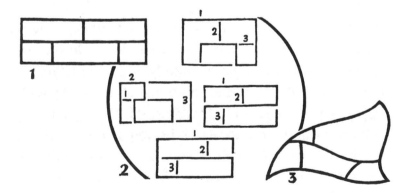

It is impossible to draw the five bricks with three pencil strokes, and there is a simple way to prove it.

When three line segments come together at a point, as shown in Figure 4, it is obvious that the point must mark the end of at least one stroke. It could also be the end of three strokes, but this does not concern us. We are interested only in the fact that at least one line must end at point P in the illustration.

4

Count the number of points, in Figure 1, showing the bricks, where three line segments meet. There are eight such points. Each must mark the end of at least one stroke, so the entire figure contains at least eight ends of strokes. No single stroke can have more than two ends, therefore the figure cannot be drawn with less than four strokes.

This is a simple example of what mathematicians call an *impossibility proof*. Very often in the history of mathematics a great deal of time is wasted in trying to solve a problem, like the trisecting of an angle with only a compass and straightedge, that has no solution. So it is very important to search for impossibility proofs. Another excellent example of such a proof will be found in the five-tetrominoes puzzle in the next section.

OUTSIDE OR INSIDE?

ONE OF THE fundamental theorems in topology is called the Jordan curve theorem. (It is named after the French mathematician Camille Jordan.) This theorem states that any *simple closed curve* (a curve that is joined at the ends and does not cross itself) divides the surface of a plane into two regions—an outside and an inside (Figure 1). The theorem seems quite obvious, but as a matter of fact it is rather difficult to prove.

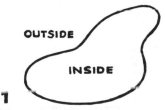

OUTSIDE

INSIDE

1

If we draw a simple closed curve that is very twisty, such as the one shown in Figure 2, it is not easy to say at once whether a certain spot, such as the one marked by the small cross, is inside or outside. Of course we can find out by tracing the region containing this spot to the edge of the curve to see if it does or does not lead outside.

2

Figure 3 shows only a small interior portion of a simple closed curve. The rest of the curve, on all four sides, is hidden from view by sheets of paper, so there is no way to trace any of the visible regions out to the boundary. We are told that the region marked A is inside the curve.

Is region B inside or outside, and how do you know?

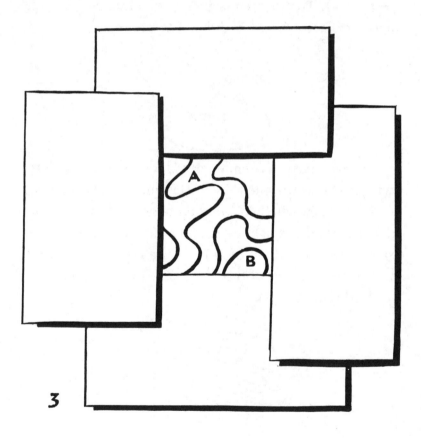

3

Region B is inside.

This can be said because of another interesting theorem about simple closed curves. All "inside" regions of such a curve are separated from each other by an *even* number of lines. The same is true of all "outside" regions. And any inside region is separated from any outside region by an *odd* number of lines. Zero is considered an even number, so if there are no lines between two regions, then of course they will be part of the same "side," and our theorem still holds.

When we pass from any part of region A to any part of region B, along any path, we cross an even number of lines. In Figure 4 one such path is shown by the dotted line. As you can see, it crosses four lines, an even number. So we can say with certainty that no matter what the rest of this curve looks like, region B is also inside!

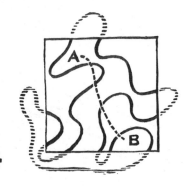

4

THE TWO KNOTS

MANY PEOPLE TODAY know what a Moebius strip is. It is a strip of paper that is given one half-twist before the ends are pasted together as shown in Figure 1. It has only one side and one edge.

Many people know also that if you try to cut a Moebius strip in half, cutting lengthwise down the middle of the strip, it will not make two strips as you would expect. It opens up into one large strip. And if you start cutting a third of the way from the side, you will cut twice around the strip to produce one large strip that has a smaller one linked through it.

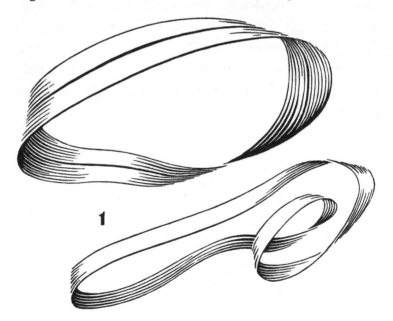

1

If you give the strip two half-twists before you paste together the ends (Figure 2), cutting down the middle produces two strips of the same size, but linked. What happens if you cut a strip that has three half-twists? (Figure 3.) This time you get one large strip that is tied in a knot! (Figure 4.)

There are two ways to make a strip with three half-twists. We can twist the strip clockwise or we can twist it counterclockwise. In both cases, cutting the strip will form a knot.

Now for the question: are these two knots exactly alike?

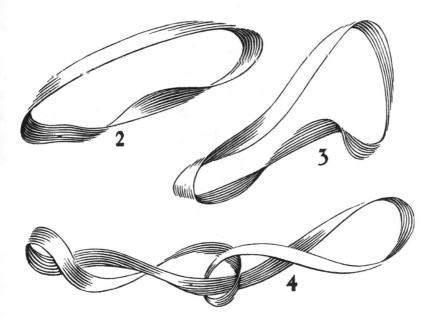

At first glance you might suppose that the two knots are alike, but if you examine them more carefully you will notice a curious difference. One knot is a mirror image of the other. No matter how we try to alter the shape of one knot, it can never be made to look exactly like the other.

Geometrical structures that are not the same as their mirror images are called *asymmetric*. When we formed the two strips, twisting one in one direction and the other in the other direction, we formed two asymmetric strips, each a mirror image of the other. This asymmetry is carried over into the asymmetry of the two knots that result from cutting.

We are so used to tying overhand knots the same way that we do not realize that there are two quite distinct ways of tying them. Perhaps left-handed people tend to tie them one way and right-handed people the other way. If so, then Sherlock Holmes would have a good way of deducing, from the way a criminal tied up his victim, whether the criminal was right- or left-handed.

REVERSING THE SWEATER

Imagine that your wrists are tied together with a piece of rope, as shown in the illustration, and that you are wearing a slipover sweater.

Is there any way that you can take off your sweater, turn it inside-out, and put it back on again? Remember, the sweater has no buttons and you are not allowed to untie or cut the rope.

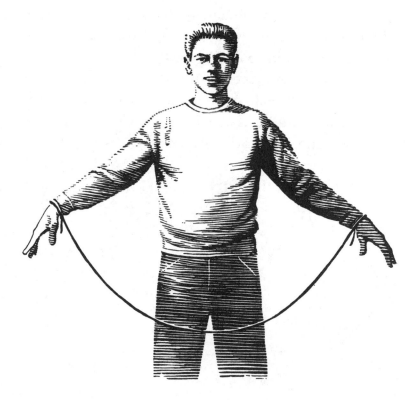

Yes, the sweater can be reversed as follows:

(1) Pull it over your head, reversing it as you do so, and allow it to hang, inside-out, on the rope as shown in Figure 1.

(2) Reverse the sweater again by pushing it through one of its sleeves. It now hangs on the rope right-side out (Figure 2).

(3) Put it on again, over your head, by going backwards through the actions by which you took it off. This reverses the sweater a third time, and puts it on your body inside-out (Figure 3).

Before trying it, see if you can visualize the process in your mind. If your sweater has a school letter sewn on the front, will this letter be touching your chest or your back after you have finished the three steps?

PART IX

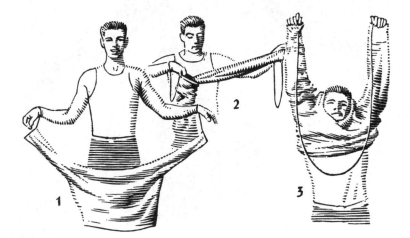

MISCELLANEOUS PUZZLES

Miscellaneous Puzzles

THERE ARE SO many different branches of mathematics that, if we included a problem taken from each of them, this book would have to be fifty times larger than it is. The next five puzzles do not fit well into any of the previous sections, but they are included here because they are especially interesting and because they introduce important mathematical ideas.

The first puzzle involves a branch of geometry called *combinatorial geometry*. It shows how to make a type of jigsaw puzzle that has interested many top flight mathematicians. The second and fifth puzzles involve *logic*. From Aristotle's day until a century ago, logic was considered part of philosophy; now it is regarded as the study of the most fundamental laws of mathematics. The third puzzle points out an amusing pitfall in a field of mathematics called *statistics*.

The fourth puzzle shows how mathematical reasoning can often increase the efficiency of work, even the work of a person having breakfast. Today, the application of mathematics to industry and warfare, to make their operations more efficient, is known as *operations research*, or O.R. It is one of the fastest growing fields of modern mathematics.

THE FIVE TETROMINOES

TRACE THE FIVE shapes shown in Figure 1 on a sheet of stiff paper or cardboard, and cut them out. Can you fit them together to make the 4 × 5 rectangle shown in Figure 2? Pieces may be turned over and placed with either side up.

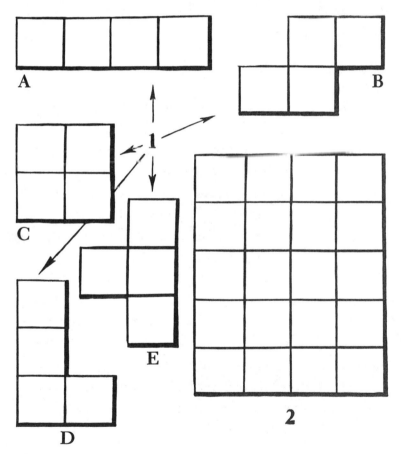

The five shapes are called *tetrominoes*. A *domino* is formed by putting together two small squares. Tetrominoes consist of four small squares joined together. Shapes made of three squares are called *trominoes*, and those made with five squares are called *pentominoes*.

The general name for all such shapes is *polyominoes*. Hundreds of interesting puzzles are based upon them.

SOLUTION

There is no way to solve this puzzle. Perhaps you convinced yourself of this by trying for a long time to form the rectangle, but without success. A mathematician, however, is never content with just suspecting that something is impossible. He wants to prove it. In this case, there is a surprisingly simple way to do so.

First color the small squares of the rectangle so that it looks like a checkerboard (Figure 3). If you try placing tetrominoes A, B, C, D on this checkerboard you will see that no matter where you place them, each must cover two black squares and two white ones. The four together, then, must always cover a total area of eight black and eight white squares.

This is not true, however, of tetromino E. It always covers three squares of one color and one of the other.

The rectangle has ten small white squares, ten

black ones. No matter where we place tetrominoes A, B, C, D, we will have to cover eight squares of each color. This will leave two black and two white squares yet to be covered by tetromino E. But E cannot cover two white and two black squares. Therefore the puzzle cannot be solved.

Figure 4 shows a figure shaped like a skyscraper, in which there are two more black squares than there are white, so our impossibility proof no longer applies. Try to form *this* figure with the five pieces. It can be done!

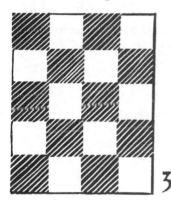

3

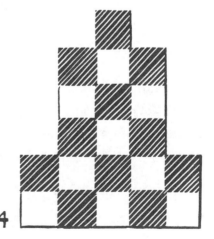

4

THE TWO TRIBES

AN ISLAND IS inhabited by two tribes. Members of one tribe always tell the truth, members of the other always lie.

A missionary met two of these natives, one tall, the other short. "Are you a truth-teller?" he asked the taller one.

"Oopf," the tall native answered.

The missionary recognized this as a native word meaning either yes or no, but he couldn't recall which. The short native spoke English, so the missionary asked him what his companion had said.

"He say 'yes,'" replied the short native, "but him big liar!"

What tribe did each native belong to?

SOLUTION

When the missionary asked the tall native if he was a truth-teller, the answer "Oopf" *has* to mean "yes." If the native is a truth-teller, he would tell the truth and answer yes; if he is a liar, he would lie and still answer yes!

So when the short native told the missionary that his companion said "yes," the short native was telling the truth. Therefore he must also have told the truth when he said his friend was a liar.

Conclusion: the tall man is a liar, the short one a truth-teller.

NO TIME FOR SCHOOL

"BUT I DON'T have *time* for school," explained Eddie to the truant officer. "I sleep eight hours a day, which adds up to about 122 days a year, assuming each day is 24 hours. There's no school on Saturday or Sunday, which amounts to 104 days a year. We have 60 days of summer vacation. I need three hours a day for meals—that's more than 45 days a year. And I need at least two hours a day for recreation—that comes to over 30 days a year."

Eddie jotted down these figures as he spoke, then he added up all the days. They came to 361.

Sleep (8 hours a day)	122
Saturdays and Sundays	104
Summer vacation	60
Meals (3 hours a day)	45
Recreation (2 hours a day)	30
Total	361 days

"You see," continued Eddie, "that leaves me only 4 days to be sick in bed, and I haven't even taken into consideration the 7 school holidays that we get every year."

The truant officer scratched his head. "Something's wrong here," he mumbled. But try as he could, he was unable to find anything inaccurate about Eddie's figures. Can you explain what is wrong?

The joker in Eddie's figures is that his time categories overlap so that the same periods of time are counted more than once. To give one example, during his vacation period of 60 days he both ate and slept. This eating and sleeping time is counted in the vacation period and also counted separately in his eating and sleeping times for the entire year.

This fallacy of overlapping categories is a very common one in statistics, especially medical statistics. You may read that in a certain community 30 per cent of the people have vitamin A deficiency, 30 per cent have vitamin B deficiency, and 30 per cent have vitamin C deficiency. If you conclude from this that only 10 per cent have no deficiency of these three vitamins, then you are guilty of the same kind of faulty reasoning that Eddie used on the truant officer. It is possible that 30 per cent of the people have deficiencies in *all three* vitamins, leaving 70 per cent of the population with no deficiencies at all.

TIME FOR TOAST

THE SMITHS OWN an old-fashioned toaster that takes only two slices of bread at a time, toasting one side of each. To toast the other sides, you have to remove the slices, turn them over, and put them back into the toaster. It takes exactly one minute for the toaster to toast one side of each piece of bread that it contains.

One morning Mrs. Smith wished to toast both sides of three slices. Mr. Smith watched over the top of his newspaper and smiled when he saw how his wife went about it. It took her four minutes.

"You could have toasted those three slices in less time, my dear," he said, "and kept down the cost of our electric bill."

Was Mr. Smith right, and if so, how could his wife have toasted those three slices in less than four minutes?

SOLUTION

It is a simple matter to toast all three slices, on both sides, in three minutes. Let's call the slices A, B, and C. Each slice has side 1 and side 2. This is the procedure:

First minute: Toast sides A1 and B1. Remove the slices, turn B around and put it back into the toaster. Put A aside and put C in the toaster.

Second minute: Toast sides B2 and C1. Remove the slices, turn C around and put back in the toaster. Put B aside (it is now toasted on both sides) and put A back in the toaster.

Third minute: Toast sides A2 and C2. All sides of all three slices are now toasted.

THE THREE NECKTIES

MR. BROWN, Mr. Green, and Mr. Black were having lunch together. One wore a brown necktie, one a green tie, one a black.

"Have you noticed," said the man with the green tie, "that although our ties have colors that match our names, not one of us has on a tie that matches his *own* name?"

"By golly, you're right!" exclaimed Mr. Brown. What color tie was each man wearing?"

SOLUTION

Mr. Brown had a black tie.

Mr. Black had a green tie.

Mr. Green had a brown tie.

Brown couldn't be wearing a brown tie, for then it would correspond to his name. He couldn't be wearing a green tie because a tie of this color is on the man who asked him a question. Therefore Brown's tie must be black.

This leaves the green and brown ties to be worn respectively by Mr. Black and Mr. Green.

PART X

TRICKY

PUZZLES

Tricky Puzzles

THE TWENTY-EIGHT PUZZLES in this section are short and easy, but each involves a funny twist of some sort that gives the answer an unexpected turn. In a way, you could call them humorous or catch problems, but I have chosen to end the book with them for a very special reason.

The reason is this: a truly creative mathematician or scientist must have a mind that is constantly on the alert for surprising, off-beat angles. Einstein, for example, the greatest scientist of recent times, would never have developed his famous theory of relativity if he had not questioned certain assumptions that for centuries other scientists had not dared to question. He tackled problems that seemed to have no solution, and solved them by finding their "catch" element—that strange hidden factor that everyone else had overlooked. Sometimes the new twist is so simple that once it is discovered other scientists wonder why *they* didn't think of it. They didn't, of course, because their minds were nailed down by habit to familiar, orthodox ways of thinking.

So limber up your brain before you try these amusing questions. They are not important mathematically, but they will teach you that in mathematics, as well as in life, things are not always what they seem!

TRICKY PUZZLES

1. Can you place ten lumps of sugar in three empty cups so that there is an odd number of lumps in each cup?

2. At the local hardware store, Jones learned that 1 would cost him 50 cents, 12 would cost $1.00, and the price of 144 was $1.50. What was Jones buying?

3. See how quickly you can jot down the digits from 9 to 1 *backwards*, then check the answer to see how carefully you followed directions.

4. How quickly can you find the product of the following numbers?

$$256 \times 3 \times 45 \times 3,901 \times 77 \times 488 \times 2,800 \times 0$$

5. Laryngitis, a Greek orator, was born on July 4, 30 B.C. He died on July 4, 30 A.D. How old was he when he died?

6. A dog and cat together weigh 27 pounds. If the dog's weight is an odd number, and if he weighs twice as much as she does, how much does each weigh?

7. After a series of experiments, a chemist discovered that it took 80 minutes for a certain chemical reaction to take place whenever he was wearing a green necktie, and the same reaction always took an hour and twenty minutes when he wore a purple tie. Can you think of why this might be so?

8. A mathematician retired at 8 o'clock one evening, set the alarm for 9 in the morning, and promptly went to sleep. When the alarm woke him up, how many hours of sleep had he had?

9. Divide 30 by ½ and add 10. What is the result?

10. A boy had five apples and ate all but three. How many were left?

11. What two whole numbers (not fractions) make the unlucky number 13 when multiplied together?

12. A reader of this book was so angry at not being able to guess the answers to all of these problems that he tore out pages 6, 7, 84, 111, and 112. How many sheets of paper did he rip out?

13. If a clock takes five seconds to strike 6 o'clock, how long will it take to strike 12 o'clock?

14. A triangle has sides of 17, 35, and 52 inches. What is its area in square inches?

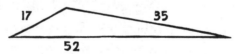

15. Can you draw four straight lines, without taking the pencil point off the paper, that will pass through all nine dots below?

16. Can you draw two straight lines, without taking the pencil from the paper, that will pass through all six baseballs in the drawing below?

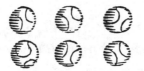

17. Each book in the set below is two inches thick. This includes the covers, which are ⅛ of an inch thick. If a bookworm starts on the first page of volume 1 and bores his way straight through the set to the last page of volume 4, how far will the worm have gone?

18. Can you circle six digits here that will add to 21?

$$
\begin{array}{ccc}
9 & 9 & 9 \\
5 & 5 & 5 \\
3 & 3 & 3 \\
1 & 1 & 1
\end{array}
$$

19. Show how to cut a pancake into eight pieces with three straight cuts of the knife.

20. An anonymously written old jingle reads as follows:

> *Four jolly men sat down to play,*
> *And played all night till break of day.*
> *They played for cash and not for fun,*
> *With separate scores for every one.*
>
> *Yet when they came to square accounts,*
> *They all had made quite fair amounts!*
> *Can you this paradox explain?*
> *If no one lost, how could all gain?*

21. A ladybug crawls along a ruler from the 12-inch mark at one end to the 6-inch mark in the center. It takes her 12 seconds. Continuing on her way, she crawls from the 6-inch mark to the 1-inch mark, but this takes her only 10 seconds. Can you think of a good reason for the time difference?

22. What is the basis for the order in which these ten digits have been arranged?

8-5-4-9-1-7-6-3-2-0

23. If there are 12 one-cent stamps in a dozen, then how many two-cent stamps are in a dozen?

24. Place a penny on each of the spots in the illustration on the facing page. Can you change the position of *one*

penny only and make two straight rows, each row containing four pennies?

25. A logician found himself in a small town that had only two barbers, each with his own shop. Needing a haircut, he glanced into one shop and saw at once that it was extremely dirty. The barber himself needed a shave, his clothes were untidy, his hair unkempt and badly cut. The other shop proved to be spick-and-span. The barber was freshly shaved, spotlessly dressed, and

his hair was neatly trimmed. The logician thought a moment, then returned to the first shop for his haircut. Why?

26. When their blind dates arrived to take them to a football game, Katy and Susan were astonished to see that the two young men looked exactly alike.

"Yes, we're brothers," one of them explained. "We were born on the same day in the same year, and we have the same parents."

"But we're not twins," said the other.

Katy and Susan were puzzled. Can you explain the situation?

27. Multiplying 10 feet by 10 feet equals 100 square feet. How much is ten dollars times ten dollars?

28. When the young man paid the cashier for his breakfast, she noticed that he had drawn a triangle on the back of the check. Underneath the triangle he had written: $13 \times 2 = 26$.

The cashier smiled. "I see you are a sailor," she said. How did the cashier know that he was a sailor?

SOLUTIONS

1. There are fifteen different solutions to this problem, but all involve the same gimmick. For example: put seven lumps in one cup, two in another, one in a third. Now place the last cup in the second one. The second cup will then contain three lumps!

2. Jones was buying house numbers.

3. The digits from 9 to 1 backwards are:

$$1\text{-}2\text{-}3\text{-}4\text{-}5\text{-}6\text{-}7\text{-}8\text{-}9$$

4. Did you notice that zero at the end before you started multiplying? It tells you at once that the final answer must be zero!

5. Laryngitis was 59 years old (there was no year zero).

6. The dog, a little Pomeranian named Henrietta, weighs 9 pounds, and the huge tomcat tips the scale at 18. If you assumed the dog was a "he" and the cat a "she," you probably gave up on this one.

7. There is nothing to explain because 80 minutes is the same as an hour and twenty minutes.

8. The mathematician had only one hour of sleep. The alarm woke him up at 9 o'clock that evening.

9. Thirty divided by ½ is 60, so when you add 10 the final answer is 70.

10. Three apples were left.

11. $13 \times 1 = 13$.

12. He ripped out only four sheets because pages 111 and 112 are two sides of the same sheet.

13. The clock will take 11 seconds to strike 12 o'clock. There is one second between each stroke.

14. A "triangle" with the sides given would be a straight line (mathematicians sometimes call this a "degenerate triangle"), so it would have no area at all. True, a triangle was shown in the illustration, but that was just to throw you off; it could not possibly have sides of the lengths indicated.

15.

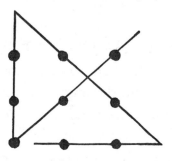

16. Because the baseballs are large spots, all of them can be crossed by drawing two lines that meet far to the right as shown.

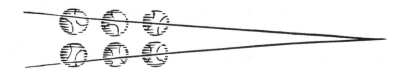

17. The first page of volume 1 is on the *right* side of the book when the volumes are standing on the shelf, and the last page of volume 4 is on the *left* side of the book. The worm therefore has only to bore through the cover of volume 1, all the way through volumes 2 and 3, and the cover of volume 4, making a total distance of 4¼ inches.

18. Turn the page upside down and circle three 6's and three 1's.

19. Two cuts at right angles will cut the pancake into four pieces. Pile them up in one pile and cut them all in half with the third cut to make eight pieces.

20. The four jolly men were four jolly jazz musicians.

21. The ladybug moves at a constant speed of one inch every two seconds. Did it occur to you that the distance from the center of the ruler to the one-inch mark is only *five* inches?

22. The digits are arranged so that their names are in alphabetical order.

23. Twelve.

24. Pick up the lowest penny and put it *on top* of the penny at the corner.

25. Since there were only two barbers in town, each must have cut the other's hair. The logician picked the barber who had given his rival the best haircut.

26. The two boys belonged to a set of triplets.

27. The question is meaningless. Dollars can be added to dollars, or subtracted from dollars, but not multiplied or divided by anything but a pure number.

28. The young man was wearing a sailor suit!

Suggestions for Further Reading

After you have finished this book, you may wish to go on to read other books that contain mathematical puzzles. Those listed below are all in print and easily obtained through any bookstore or at your library.

EASY PUZZLES

The Arrow Book of Brain Teasers, by Martin Gardner. New York: Tab Books, 1959, 64 pages. Amusing puzzles for pre-teeners, in an inexpensive paperback.

Encyclopedia of Puzzles and Pastimes, edited by Clark Kinnaird. New York: Grosset and Dunlap, Inc., 1946, 431 pages. A mammoth collection of puzzles of all types, most of them reprinted from the newspaper puzzle page syndicated by King Features.

Figures for Fun, by Yakov Perelman. Moscow: Foreign Languages Publishing House, 1957. An English translation of a book by Russia's leading puzzlist.

NOT-SO-EASY PUZZLES

The Canterbury Puzzles, by H. E. Dudeney. New York: Dover Publications, Inc., 1959, 255 pages. A paperback reissue of the first puzzle book by England's greatest puzzlist.

Amusements in Mathematics, by H. E. Dudeney. New York: Dover Publications, Inc., 1959, 258 pages. A paperback reissue of the author's second collection of puzzles
    Puzzle-Math, by George Gamow and Marvin Stern. New York: The Viking Press, 1958, 128 pages. A splendid collection of amusing new brain teasers.
    The Scientific American Book of Mathematical Puzzles and Diversions, by Martin Gardner. New York: Simon and Schuster, Inc., 1959, 178 pages. The first sixteen of the author's puzzle columns in *Scientific American,* with new material added.
    Best Mathematical Puzzles of Sam Loyd, edited by Martin Gardner. New York: Dover Publications, Inc., 1959, 167 pages. A paperback selection from the fabulous, long out-of-print *Cyclopedia of Puzzles* by Loyd, the greatest of U.S. puzzlists.
    Mathematical Puzzles for Beginners and Enthusiasts, by Geoffrey Mott-Smith. New York: Dover Publications, Inc., 1954, 256 pages. An excellent paperback selection by one of the nation's leading experts on games and puzzles.

A CATALOG OF SELECTED
DOVER BOOKS
IN ALL FIELDS OF INTEREST

A CATALOG OF SELECTED DOVER
BOOKS IN ALL FIELDS OF INTEREST

CONCERNING THE SPIRITUAL IN ART, Wassily Kandinsky. Pioneering work by father of abstract art. Thoughts on color theory, nature of art. Analysis of earlier masters. 12 illustrations. 80pp. of text. 5⅜ x 8½. 23411-8

ANIMALS: 1,419 Copyright-Free Illustrations of Mammals, Birds, Fish, Insects, etc., Jim Harter (ed.). Clear wood engravings present, in extremely lifelike poses, over 1,000 species of animals. One of the most extensive pictorial sourcebooks of its kind. Captions. Index. 284pp. 9 x 12. 23766-4

CELTIC ART: The Methods of Construction, George Bain. Simple geometric techniques for making Celtic interlacements, spirals, Kells-type initials, animals, humans, etc. Over 500 illustrations. 160pp. 9 x 12. (Available in U.S. only.) 22923-8

AN ATLAS OF ANATOMY FOR ARTISTS, Fritz Schider. Most thorough reference work on art anatomy in the world. Hundreds of illustrations, including selections from works by Vesalius, Leonardo, Goya, Ingres, Michelangelo, others. 593 illustrations. 192pp. 7⅛ x 10¼. 20241-0

CELTIC HAND STROKE-BY-STROKE (Irish Half-Uncial from "The Book of Kells"): An Arthur Baker Calligraphy Manual, Arthur Baker. Complete guide to creating each letter of the alphabet in distinctive Celtic manner. Covers hand position, strokes, pens, inks, paper, more. Illustrated. 48pp. 8¼ x 11. 24336-2

EASY ORIGAMI, John Montroll. Charming collection of 32 projects (hat, cup, pelican, piano, swan, many more) specially designed for the novice origami hobbyist. Clearly illustrated easy-to-follow instructions insure that even beginning papercrafters will achieve successful results. 48pp. 8¼ x 11. 27298-2

THE COMPLETE BOOK OF BIRDHOUSE CONSTRUCTION FOR WOODWORKERS, Scott D. Campbell. Detailed instructions, illustrations, tables. Also data on bird habitat and instinct patterns. Bibliography. 3 tables. 63 illustrations in 15 figures. 48pp. 5¼ x 8½. 24407-5

BLOOMINGDALE'S ILLUSTRATED 1886 CATALOG: Fashions, Dry Goods and Housewares, Bloomingdale Brothers. Famed merchants' extremely rare catalog depicting about 1,700 products: clothing, housewares, firearms, dry goods, jewelry, more. Invaluable for dating, identifying vintage items. Also, copyright-free graphics for artists, designers. Co-published with Henry Ford Museum & Greenfield Village. 160pp. 8¼ x 11. 25780-0

HISTORIC COSTUME IN PICTURES, Braun & Schneider. Over 1,450 costumed figures in clearly detailed engravings—from dawn of civilization to end of 19th century. Captions. Many folk costumes. 256pp. 8⅜ x 11¾. 23150-X

STICKLEY CRAFTSMAN FURNITURE CATALOGS, Gustav Stickley and L. & J. G. Stickley. Beautiful, functional furniture in two authentic catalogs from 1910. 594 illustrations, including 277 photos, show settles, rockers, armchairs, reclining chairs, bookcases, desks, tables. 183pp. 6½ x 9¼. 23838-5

AMERICAN LOCOMOTIVES IN HISTORIC PHOTOGRAPHS: 1858 to 1949, Ron Ziel (ed.). A rare collection of 126 meticulously detailed official photographs, called "builder portraits," of American locomotives that majestically chronicle the rise of steam locomotive power in America. Introduction. Detailed captions. xi+ 129pp. 9 x 12. 27393-8

AMERICA'S LIGHTHOUSES: An Illustrated History, Francis Ross Holland, Jr. Delightfully written, profusely illustrated fact-filled survey of over 200 American light-houses since 1716. History, anecdotes, technological advances, more. 240pp. 8 x 10¾. 25576-X

TOWARDS A NEW ARCHITECTURE, Le Corbusier. Pioneering manifesto by founder of "International School." Technical and aesthetic theories, views of industry, eco-nomics, relation of form to function, "mass-production split" and much more. Profusely illustrated. 320pp. 6⅛ x 9¼. (Available in U.S. only.) 25023-7

HOW THE OTHER HALF LIVES, Jacob Riis. Famous journalistic record, expos-ing poverty and degradation of New York slums around 1900, by major social reformer. 100 striking and influential photographs. 233pp. 10 x 7⅞. 22012-5

FRUIT KEY AND TWIG KEY TO TREES AND SHRUBS, William M. Harlow. One of the handiest and most widely used identification aids. Fruit key covers 120 deciduous and evergreen species; twig key 160 deciduous species. Easily used. Over 300 photographs. 126pp. 5⅜ x 8½. 20511-8

COMMON BIRD SONGS, Dr. Donald J. Borror. Songs of 60 most common U.S. birds: robins, sparrows, cardinals, bluejays, finches, more–arranged in order of increasing complexity. Up to 9 variations of songs of each species.
Cassette and manual 99911-4

ORCHIDS AS HOUSE PLANTS, Rebecca Tyson Northen. Grow cattleyas and many other kinds of orchids–in a window, in a case, or under artificial light. 63 illus-trations. 148pp. 5⅜ x 8½. 23261-1

MONSTER MAZES, Dave Phillips. Masterful mazes at four levels of difficulty. Avoid deadly perils and evil creatures to find magical treasures. Solutions for all 32 exciting illustrated puzzles. 48pp. 8¼ x 11. 26005-4

MOZART'S DON GIOVANNI (DOVER OPERA LIBRETTO SERIES), Wolfgang Amadeus Mozart. Introduced and translated by Ellen H. Bleiler. Standard Italian libretto, with complete English translation. Convenient and thoroughly portable–an ideal companion for reading along with a recording or the performance itself. Introduction. List of characters. Plot summary. 121pp. 5¼ x 8½. 24944-1

TECHNICAL MANUAL AND DICTIONARY OF CLASSICAL BALLET, Gail Grant. Defines, explains, comments on steps, movements, poses and concepts. 15-page pictorial section. Basic book for student, viewer. 127pp. 5⅜ x 8½. 21843-0

THE CLARINET AND CLARINET PLAYING, David Pino. Lively, comprehensive work features suggestions about technique, musicianship, and musical interpretation, as well as guidelines for teaching, making your own reeds, and preparing for public performance. Includes an intriguing look at clarinet history. "A godsend," *The Clarinet*, Journal of the International Clarinet Society. Appendixes. 7 illus. 320pp. 5⅜ x 8½. 40270-3

HOLLYWOOD GLAMOR PORTRAITS, John Kobal (ed.). 145 photos from 1926-49. Harlow, Gable, Bogart, Bacall; 94 stars in all. Full background on photographers, technical aspects. 160pp. 8⅜ x 11¼. 23352-9

THE ANNOTATED CASEY AT THE BAT: A Collection of Ballads about the Mighty Casey/Third, Revised Edition, Martin Gardner (ed.). Amusing sequels and parodies of one of America's best-loved poems: Casey's Revenge, Why Casey Whiffed, Casey's Sister at the Bat, others. 256pp. 5⅜ x 8½. 28598-7

THE RAVEN AND OTHER FAVORITE POEMS, Edgar Allan Poe. Over 40 of the author's most memorable poems: "The Bells," "Ulalume," "Israfel," "To Helen," "The Conqueror Worm," "Eldorado," "Annabel Lee," many more. Alphabetic lists of titles and first lines. 64pp. 5$^{3}/_{16}$ x 8¼. 26685-0

PERSONAL MEMOIRS OF U. S. GRANT, Ulysses Simpson Grant. Intelligent, deeply moving firsthand account of Civil War campaigns, considered by many the finest military memoirs ever written. Includes letters, historic photographs, maps and more. 528pp. 6⅛ x 9¼. 28587-1

ANCIENT EGYPTIAN MATERIALS AND INDUSTRIES, A. Lucas and J. Harris. Fascinating, comprehensive, thoroughly documented text describes this ancient civilization's vast resources and the processes that incorporated them in daily life, including the use of animal products, building materials, cosmetics, perfumes and incense, fibers, glazed ware, glass and its manufacture, materials used in the mummification process, and much more. 544pp. 6⅛ x 9¼. (Available in U.S. only.) 40446-3

RUSSIAN STORIES/RUSSKIE RASSKAZY: A Dual-Language Book, edited by Gleb Struve. Twelve tales by such masters as Chekhov, Tolstoy, Dostoevsky, Pushkin, others. Excellent word-for-word English translations on facing pages, plus teaching and study aids, Russian/English vocabulary, biographical/critical introductions, more. 416pp. 5⅜ x 8½. 26244-8

PHILADELPHIA THEN AND NOW: 60 Sites Photographed in the Past and Present, Kenneth Finkel and Susan Oyama. Rare photographs of City Hall, Logan Square, Independence Hall, Betsy Ross House, other landmarks juxtaposed with contemporary views. Captures changing face of historic city. Introduction. Captions. 128pp. 8¼ x 11. 25790-8

AIA ARCHITECTURAL GUIDE TO NASSAU AND SUFFOLK COUNTIES, LONG ISLAND, The American Institute of Architects, Long Island Chapter, and the Society for the Preservation of Long Island Antiquities. Comprehensive, well-researched and generously illustrated volume brings to life over three centuries of Long Island's great architectural heritage. More than 240 photographs with authoritative, extensively detailed captions. 176pp. 8¼ x 11. 26946-9

NORTH AMERICAN INDIAN LIFE: Customs and Traditions of 23 Tribes, Elsie Clews Parsons (ed.). 27 fictionalized essays by noted anthropologists examine religion, customs, government, additional facets of life among the Winnebago, Crow, Zuni, Eskimo, other tribes. 480pp. 6⅛ x 9¼. 27377-6

FRANK LLOYD WRIGHT'S DANA HOUSE, Donald Hoffmann. Pictorial essay of residential masterpiece with over 160 interior and exterior photos, plans, elevations, sketches and studies. 128pp. 9¼ x 10¾. 29120-0

THE MALE AND FEMALE FIGURE IN MOTION: 60 Classic Photographic Sequences, Eadweard Muybridge. 60 true-action photographs of men and women walking, running, climbing, bending, turning, etc., reproduced from rare 19th-century masterpiece. vi + 121pp. 9 x 12. 24745-7

1001 QUESTIONS ANSWERED ABOUT THE SEASHORE, N. J. Berrill and Jacquelyn Berrill. Queries answered about dolphins, sea snails, sponges, starfish, fishes, shore birds, many others. Covers appearance, breeding, growth, feeding, much more. 305pp. 5¼ x 8¼. 23366-9

ATTRACTING BIRDS TO YOUR YARD, William J. Weber. Easy-to follow guide offers advice on how to attract the greatest diversity of birds: birdhouses, feeders, water and waterers, much more. 96pp. 5³⁄₁₆ x 8¼. 28927-3

MEDICINAL AND OTHER USES OF NORTH AMERICAN PLANTS: A Historical Survey with Special Reference to the Eastern Indian Tribes, Charlotte Erichsen-Brown. Chronological historical citations document 500 years of usage of plants, trees, shrubs native to eastern Canada, northeastern U.S. Also complete identifying information. 343 illustrations. 544pp. 6½ x 9¼. 25951-X

STORYBOOK MAZES, Dave Phillips. 23 stories and mazes on two-page spreads: Wizard of Oz, Treasure Island, Robin Hood, etc. Solutions. 64pp. 8¼ x 11. 23628-5

AMERICAN NEGRO SONGS: 230 Folk Songs and Spirituals, Religious and Secular, John W. Work. This authoritative study traces the African influences of songs sung and played by black Americans at work, in church, and as entertainment. The author discusses the lyric significance of such songs as "Swing Low, Sweet Chariot," "John Henry," and others and offers the words and music for 230 songs. Bibliography. Index of Song Titles. 272pp. 6½ x 9¼. 40271-1

MOVIE-STAR PORTRAITS OF THE FORTIES, John Kobal (ed.). 163 glamor, studio photos of 106 stars of the 1940s: Rita Hayworth, Ava Gardner, Marlon Brando, Clark Gable, many more. 176pp. 8⅜ x 11¼. 23546-7

BENCHLEY LOST AND FOUND, Robert Benchley. Finest humor from early 30s, about pet peeves, child psychologists, post office and others. Mostly unavailable elsewhere. 73 illustrations by Peter Arno and others. 183pp. 5⅜ x 8½. 22410-4

YEKL and THE IMPORTED BRIDEGROOM AND OTHER STORIES OF YIDDISH NEW YORK, Abraham Cahan. Film Hester Street based on *Yekl* (1896). Novel, other stories among first about Jewish immigrants on N.Y.'s East Side. 240pp. 5⅜ x 8½. 22427-9

SELECTED POEMS, Walt Whitman. Generous sampling from *Leaves of Grass*. Twenty-four poems include "I Hear America Singing," "Song of the Open Road," "I Sing the Body Electric," "When Lilacs Last in the Dooryard Bloom'd," "O Captain! My Captain!"—all reprinted from an authoritative edition. Lists of titles and first lines. 128pp. 5³⁄₁₆ x 8¼. 26878-0

THE BEST TALES OF HOFFMANN, E. T. A. Hoffmann. 10 of Hoffmann's most important stories: "Nutcracker and the King of Mice," "The Golden Flowerpot," etc. 458pp. 5⅜ x 8½. 21793-0

FROM FETISH TO GOD IN ANCIENT EGYPT, E. A. Wallis Budge. Rich detailed survey of Egyptian conception of "God" and gods, magic, cult of animals, Osiris, more. Also, superb English translations of hymns and legends. 240 illustrations. 545pp. 5⅜ x 8½. 25803-3

FRENCH STORIES/CONTES FRANÇAIS: A Dual-Language Book, Wallace Fowlie. Ten stories by French masters, Voltaire to Camus: "Micromegas" by Voltaire; "The Atheist's Mass" by Balzac; "Minuet" by de Maupassant; "The Guest" by Camus, six more. Excellent English translations on facing pages. Also French-English vocabulary list, exercises, more. 352pp. 5⅜ x 8½. 26443-2

CHICAGO AT THE TURN OF THE CENTURY IN PHOTOGRAPHS: 122 Historic Views from the Collections of the Chicago Historical Society, Larry A. Viskochil. Rare large-format prints offer detailed views of City Hall, State Street, the Loop, Hull House, Union Station, many other landmarks, circa 1904-1913. Introduction. Captions. Maps. 144pp. 9⅜ x 12¼. 24656-6

OLD BROOKLYN IN EARLY PHOTOGRAPHS, 1865-1929, William Lee Younger. Luna Park, Gravesend race track, construction of Grand Army Plaza, moving of Hotel Brighton, etc. 157 previously unpublished photographs. 165pp. 8⅞ x 11¾. 23587-4

THE MYTHS OF THE NORTH AMERICAN INDIANS, Lewis Spence. Rich anthology of the myths and legends of the Algonquins, Iroquois, Pawnees and Sioux, prefaced by an extensive historical and ethnological commentary. 36 illustrations. 480pp. 5⅜ x 8½. 25967-6

AN ENCYCLOPEDIA OF BATTLES: Accounts of Over 1,560 Battles from 1479 B.C. to the Present, David Eggenberger. Essential details of every major battle in recorded history from the first battle of Megiddo in 1479 B.C. to Grenada in 1984. List of Battle Maps. New Appendix covering the years 1967-1984. Index. 99 illustrations. 544pp. 6½ x 9¼. 24913-1

SAILING ALONE AROUND THE WORLD, Captain Joshua Slocum. First man to sail around the world, alone, in small boat. One of great feats of seamanship told in delightful manner. 67 illustrations. 294pp. 5⅜ x 8½. 20326-3

ANARCHISM AND OTHER ESSAYS, Emma Goldman. Powerful, penetrating, prophetic essays on direct action, role of minorities, prison reform, puritan hypocrisy, violence, etc. 271pp. 5⅜ x 8½. 22484-8

MYTHS OF THE HINDUS AND BUDDHISTS, Ananda K. Coomaraswamy and Sister Nivedita. Great stories of the epics; deeds of Krishna, Shiva, taken from puranas, Vedas, folk tales; etc. 32 illustrations. 400pp. 5⅜ x 8½. 21759-0

THE TRAUMA OF BIRTH, Otto Rank. Rank's controversial thesis that anxiety neurosis is caused by profound psychological trauma which occurs at birth. 256pp. 5⅜ x 8½. 27974-X

A THEOLOGICO-POLITICAL TREATISE, Benedict Spinoza. Also contains unfinished Political Treatise. Great classic on religious liberty, theory of government on common consent. R. Elwes translation. Total of 421pp. 5⅜ x 8½. 20249-6

CATALOG OF DOVER BOOKS

MY BONDAGE AND MY FREEDOM, Frederick Douglass. Born a slave, Douglass became outspoken force in antislavery movement. The best of Douglass' autobiographies. Graphic description of slave life. 464pp. 5⅜ x 8½. 22457-0

FOLLOWING THE EQUATOR: A Journey Around the World, Mark Twain. Fascinating humorous account of 1897 voyage to Hawaii, Australia, India, New Zealand, etc. Ironic, bemused reports on peoples, customs, climate, flora and fauna, politics, much more. 197 illustrations. 720pp. 5⅜ x 8½. 26113-1

THE PEOPLE CALLED SHAKERS, Edward D. Andrews. Definitive study of Shakers: origins, beliefs, practices, dances, social organization, furniture and crafts, etc. 33 illustrations. 351pp. 5⅜ x 8½. 21081-2

THE MYTHS OF GREECE AND ROME, H. A. Guerber. A classic of mythology, generously illustrated, long prized for its simple, graphic, accurate retelling of the principal myths of Greece and Rome, and for its commentary on their origins and significance. With 64 illustrations by Michelangelo, Raphael, Titian, Rubens, Canova, Bernini and others. 480pp. 5⅜ x 8½. 27584-1

PSYCHOLOGY OF MUSIC, Carl E. Seashore. Classic work discusses music as a medium from psychological viewpoint. Clear treatment of physical acoustics, auditory apparatus, sound perception, development of musical skills, nature of musical feeling, host of other topics. 88 figures. 408pp. 5⅜ x 8½. 21851-1

THE PHILOSOPHY OF HISTORY, Georg W. Hegel. Great classic of Western thought develops concept that history is not chance but rational process, the evolution of freedom. 457pp. 5⅜ x 8½. 20112-0

THE BOOK OF TEA, Kakuzo Okakura. Minor classic of the Orient: entertaining, charming explanation, interpretation of traditional Japanese culture in terms of tea ceremony. 94pp. 5⅜ x 8½. 20070-1

LIFE IN ANCIENT EGYPT, Adolf Erman. Fullest, most thorough, detailed older account with much not in more recent books, domestic life, religion, magic, medicine, commerce, much more. Many illustrations reproduce tomb paintings, carvings, hieroglyphs, etc. 597pp. 5⅜ x 8½. 22632-8

SUNDIALS, Their Theory and Construction, Albert Waugh. Far and away the best, most thorough coverage of ideas, mathematics concerned, types, construction, adjusting anywhere. Simple, nontechnical treatment allows even children to build several of these dials. Over 100 illustrations. 230pp. 5⅜ x 8½. 22947-5

THEORETICAL HYDRODYNAMICS, L. M. Milne-Thomson. Classic exposition of the mathematical theory of fluid motion, applicable to both hydrodynamics and aerodynamics. Over 600 exercises. 768pp. 6⅛ x 9¼. 68970-0

SONGS OF EXPERIENCE: Facsimile Reproduction with 26 Plates in Full Color, William Blake. 26 full-color plates from a rare 1826 edition. Includes "The Tyger," "London," "Holy Thursday," and other poems. Printed text of poems. 48pp. 5¼ x 7. 24636-1

OLD-TIME VIGNETTES IN FULL COLOR, Carol Belanger Grafton (ed.). Over 390 charming, often sentimental illustrations, selected from archives of Victorian graphics—pretty women posing, children playing, food, flowers, kittens and puppies, smiling cherubs, birds and butterflies, much more. All copyright-free. 48pp. 9¼ x 12¼. 27269-9

PERSPECTIVE FOR ARTISTS, Rex Vicat Cole. Depth, perspective of sky and sea, shadows, much more, not usually covered. 391 diagrams, 81 reproductions of drawings and paintings. 279pp. 5⅜ x 8½. 22487-2

DRAWING THE LIVING FIGURE, Joseph Sheppard. Innovative approach to artistic anatomy focuses on specifics of surface anatomy, rather than muscles and bones. Over 170 drawings of live models in front, back and side views, and in widely varying poses. Accompanying diagrams. 177 illustrations. Introduction. Index. 144pp. 8⅜ x11¼. 26723-7

GOTHIC AND OLD ENGLISH ALPHABETS: 100 Complete Fonts, Dan X. Solo. Add power, elegance to posters, signs, other graphics with 100 stunning copyright-free alphabets: Blackstone, Dolbey, Germania, 97 more–including many lower-case, numerals, punctuation marks. 104pp. 8⅛ x 11. 24695-7

HOW TO DO BEADWORK, Mary White. Fundamental book on craft from simple projects to five-bead chains and woven works. 106 illustrations. 142pp. 5⅜ x 8. 20697-1

THE BOOK OF WOOD CARVING, Charles Marshall Sayers. Finest book for beginners discusses fundamentals and offers 34 designs. "Absolutely first rate . . . well thought out and well executed."–E. J. Tangerman. 118pp. 7¾ x 10⅝. 23654-4

ILLUSTRATED CATALOG OF CIVIL WAR MILITARY GOODS: Union Army Weapons, Insignia, Uniform Accessories, and Other Equipment, Schuyler, Hartley, and Graham. Rare, profusely illustrated 1846 catalog includes Union Army uniform and dress regulations, arms and ammunition, coats, insignia, flags, swords, rifles, etc. 226 illustrations. 160pp. 9 x 12. 24939-5

WOMEN'S FASHIONS OF THE EARLY 1900s: An Unabridged Republication of "New York Fashions, 1909," National Cloak & Suit Co. Rare catalog of mail-order fashions documents women's and children's clothing styles shortly after the turn of the century. Captions offer full descriptions, prices. Invaluable resource for fashion, costume historians. Approximately 725 illustrations. 128pp. 8⅜ x 11¼. 27276-1

THE 1912 AND 1915 GUSTAV STICKLEY FURNITURE CATALOGS, Gustav Stickley. With over 200 detailed illustrations and descriptions, these two catalogs are essential reading and reference materials and identification guides for Stickley furniture. Captions cite materials, dimensions and prices. 112pp. 6½ x 9¼. 26676-1

EARLY AMERICAN LOCOMOTIVES, John H. White, Jr. Finest locomotive engravings from early 19th century: historical (1804–74), main-line (after 1870), special, foreign, etc. 147 plates. 142pp. 11⅜ x 8¼. 22772-3

THE TALL SHIPS OF TODAY IN PHOTOGRAPHS, Frank O. Braynard. Lavishly illustrated tribute to nearly 100 majestic contemporary sailing vessels: Amerigo Vespucci, Clearwater, Constitution, Eagle, Mayflower, Sea Cloud, Victory, many more. Authoritative captions provide statistics, background on each ship. 190 black-and-white photographs and illustrations. Introduction. 128pp. 8⅞ x 11¾. 27163-3

CATALOG OF DOVER BOOKS

LITTLE BOOK OF EARLY AMERICAN CRAFTS AND TRADES, Peter Stockham (ed.). 1807 children's book explains crafts and trades: baker, hatter, cooper, potter, and many others. 23 copperplate illustrations. 140pp. 4⅝ x 6. 23336-7

VICTORIAN FASHIONS AND COSTUMES FROM HARPER'S BAZAR, 1867–1898, Stella Blum (ed.). Day costumes, evening wear, sports clothes, shoes, hats, other accessories in over 1,000 detailed engravings. 320pp. 9⅜ x 12¼. 22990-4

GUSTAV STICKLEY, THE CRAFTSMAN, Mary Ann Smith. Superb study surveys broad scope of Stickley's achievement, especially in architecture. Design philosophy, rise and fall of the Craftsman empire, descriptions and floor plans for many Craftsman houses, more. 86 black-and-white halftones. 31 line illustrations. Introduction 208pp. 6½ x 9¼. 27210-9

THE LONG ISLAND RAIL ROAD IN EARLY PHOTOGRAPHS, Ron Ziel. Over 220 rare photos, informative text document origin (1844) and development of rail service on Long Island. Vintage views of early trains, locomotives, stations, passengers, crews, much more. Captions. 8⅞ x 11¾. 26301-0

VOYAGE OF THE LIBERDADE, Joshua Slocum. Great 19th-century mariner's thrilling, first-hand account of the wreck of his ship off South America, the 35-foot boat he built from the wreckage, and its remarkable voyage home. 128pp. 5⅜ x 8½. 40022-0

TEN BOOKS ON ARCHITECTURE, Vitruvius. The most important book ever written on architecture. Early Roman aesthetics, technology, classical orders, site selection, all other aspects. Morgan translation. 331pp. 5⅜ x 8½. 20645-9

THE HUMAN FIGURE IN MOTION, Eadweard Muybridge. More than 4,500 stopped-action photos, in action series, showing undraped men, women, children jumping, lying down, throwing, sitting, wrestling, carrying, etc. 390pp. 7⅞ x 10⅝. 20204-6 Clothbd.

TREES OF THE EASTERN AND CENTRAL UNITED STATES AND CANADA, William M. Harlow. Best one-volume guide to 140 trees. Full descriptions, woodlore, range, etc. Over 600 illustrations. Handy size. 288pp. 4½ x 6⅜. 20395-6

SONGS OF WESTERN BIRDS, Dr. Donald J. Borror. Complete song and call repertoire of 60 western species, including flycatchers, juncoes, cactus wrens, many more—includes fully illustrated booklet. Cassette and manual 99913-0

GROWING AND USING HERBS AND SPICES, Milo Miloradovich. Versatile handbook provides all the information needed for cultivation and use of all the herbs and spices available in North America. 4 illustrations. Index. Glossary. 236pp. 5⅜ x 8½. 25058-X

BIG BOOK OF MAZES AND LABYRINTHS, Walter Shepherd. 50 mazes and labyrinths in all—classical, solid, ripple, and more—in one great volume. Perfect inexpensive puzzler for clever youngsters. Full solutions. 112pp. 8⅛ x 11. 22951-3

PIANO TUNING, J. Cree Fischer. Clearest, best book for beginner, amateur. Simple repairs, raising dropped notes, tuning by easy method of flattened fifths. No previous skills needed. 4 illustrations. 201pp. 5⅜ x 8½. 23267-0

HINTS TO SINGERS, Lillian Nordica. Selecting the right teacher, developing confidence, overcoming stage fright, and many other important skills receive thoughtful discussion in this indispensible guide, written by a world-famous diva of four decades' experience. 96pp. 5⅜ x 8½. 40094-8

THE COMPLETE NONSENSE OF EDWARD LEAR, Edward Lear. All nonsense limericks, zany alphabets, Owl and Pussycat, songs, nonsense botany, etc., illustrated by Lear. Total of 320pp. 5⅜ x 8½. (Available in U.S. only.) 20167-8

VICTORIAN PARLOUR POETRY: An Annotated Anthology, Michael R. Turner. 117 gems by Longfellow, Tennyson, Browning, many lesser-known poets. "The Village Blacksmith," "Curfew Must Not Ring Tonight," "Only a Baby Small," dozens more, often difficult to find elsewhere. Index of poets, titles, first lines. xxiii + 325pp. 5⅝ x 8¼. 27044-0

DUBLINERS, James Joyce. Fifteen stories offer vivid, tightly focused observations of the lives of Dublin's poorer classes. At least one, "The Dead," is considered a masterpiece. Reprinted complete and unabridged from standard edition. 160pp. 5³⁄₁₆ x 8¼. 26870-5

GREAT WEIRD TALES: 14 Stories by Lovecraft, Blackwood, Machen and Others, S. T. Joshi (ed.). 14 spellbinding tales, including "The Sin Eater," by Fiona McLeod, "The Eye Above the Mantel," by Frank Belknap Long, as well as renowned works by R. H. Barlow, Lord Dunsany, Arthur Machen, W. C. Morrow and eight other masters of the genre. 256pp. 5⅜ x 8½. (Available in U.S. only.) 40436-6

THE BOOK OF THE SACRED MAGIC OF ABRAMELIN THE MAGE, translated by S. MacGregor Mathers. Medieval manuscript of ceremonial magic. Basic document in Aleister Crowley, Golden Dawn groups. 268pp. 5⅜ x 8½. 23211-5

NEW RUSSIAN-ENGLISH AND ENGLISH-RUSSIAN DICTIONARY, M. A. O'Brien. This is a remarkably handy Russian dictionary, containing a surprising amount of information, including over 70,000 entries. 366pp. 4½ x 6⅛. 20208-9

HISTORIC HOMES OF THE AMERICAN PRESIDENTS, Second, Revised Edition, Irvin Haas. A traveler's guide to American Presidential homes, most open to the public, depicting and describing homes occupied by every American President from George Washington to George Bush. With visiting hours, admission charges, travel routes. 175 photographs. Index. 160pp. 8¼ x 11. 26751-2

NEW YORK IN THE FORTIES, Andreas Feininger. 162 brilliant photographs by the well-known photographer, formerly with *Life* magazine. Commuters, shoppers, Times Square at night, much else from city at its peak. Captions by John von Hartz. 181pp. 9¼ x 10¾. 23585-8

INDIAN SIGN LANGUAGE, William Tomkins. Over 525 signs developed by Sioux and other tribes. Written instructions and diagrams. Also 290 pictographs. 111pp. 6⅛ x 9¼. 22029-X

ANATOMY: A Complete Guide for Artists, Joseph Sheppard. A master of figure drawing shows artists how to render human anatomy convincingly. Over 460 illustrations. 224pp. 8⅜ x 11¼. 27279-6

MEDIEVAL CALLIGRAPHY: Its History and Technique, Marc Drogin. Spirited history, comprehensive instruction manual covers 13 styles (ca. 4th century through 15th). Excellent photographs; directions for duplicating medieval techniques with modern tools. 224pp. 8⅜ x 11¼. 26142-5

DRIED FLOWERS: How to Prepare Them, Sarah Whitlock and Martha Rankin. Complete instructions on how to use silica gel, meal and borax, perlite aggregate, sand and borax, glycerine and water to create attractive permanent flower arrangements. 12 illustrations. 32pp. 5⅜ x 8½. 21802-3

EASY-TO-MAKE BIRD FEEDERS FOR WOODWORKERS, Scott D. Campbell. Detailed, simple-to-use guide for designing, constructing, caring for and using feeders. Text, illustrations for 12 classic and contemporary designs. 96pp. 5⅜ x 8½. 25847-5

SCOTTISH WONDER TALES FROM MYTH AND LEGEND, Donald A. Mackenzie. 16 lively tales tell of giants rumbling down mountainsides, of a magic wand that turns stone pillars into warriors, of gods and goddesses, evil hags, powerful forces and more. 240pp. 5⅜ x 8½. 29677-6

THE HISTORY OF UNDERCLOTHES, C. Willett Cunnington and Phyllis Cunnington. Fascinating, well-documented survey covering six centuries of English undergarments, enhanced with over 100 illustrations: 12th-century laced-up bodice, footed long drawers (1795), 19th-century bustles, 19th-century corsets for men, Victorian "bust improvers," much more. 272pp. 5⅜ x 8¼. 27124-2

ARTS AND CRAFTS FURNITURE: The Complete Brooks Catalog of 1912, Brooks Manufacturing Co. Photos and detailed descriptions of more than 150 now very collectible furniture designs from the Arts and Crafts movement depict davenports, settees, buffets, desks, tables, chairs, bedsteads, dressers and more, all built of solid, quarter-sawed oak. Invaluable for students and enthusiasts of antiques, Americana and the decorative arts. 80pp. 6½ x 9¼. 27471-3

WILBUR AND ORVILLE: A Biography of the Wright Brothers, Fred Howard. Definitive, crisply written study tells the full story of the brothers' lives and work. A vividly written biography, unparalleled in scope and color, that also captures the spirit of an extraordinary era. 560pp. 6⅛ x 9¼. 40297-5

THE ARTS OF THE SAILOR: Knotting, Splicing and Ropework, Hervey Garrett Smith. Indispensable shipboard reference covers tools, basic knots and useful hitches; handsewing and canvas work, more. Over 100 illustrations. Delightful reading for sea lovers. 256pp. 5⅜ x 8½. 26440-8

FRANK LLOYD WRIGHT'S FALLINGWATER: The House and Its History, Second, Revised Edition, Donald Hoffmann. A total revision—both in text and illustrations—of the standard document on Fallingwater, the boldest, most personal architectural statement of Wright's mature years, updated with valuable new material from the recently opened Frank Lloyd Wright Archives. "Fascinating"—*The New York Times*. 116 illustrations. 128pp. 9¼ x 10¾. 27430-6

PHOTOGRAPHIC SKETCHBOOK OF THE CIVIL WAR, Alexander Gardner. 100 photos taken on field during the Civil War. Famous shots of Manassas Harper's Ferry, Lincoln, Richmond, slave pens, etc. 244pp. 10⅝ x 8¼. 22731-6

FIVE ACRES AND INDEPENDENCE, Maurice G. Kains. Great back-to-the-land classic explains basics of self-sufficient farming. The one book to get. 95 illustrations. 397pp. 5⅜ x 8½. 20974-1

SONGS OF EASTERN BIRDS, Dr. Donald J. Borror. Songs and calls of 60 species most common to eastern U.S.: warblers, woodpeckers, flycatchers, thrushes, larks, many more in high-quality recording. Cassette and manual 99912-2

A MODERN HERBAL, Margaret Grieve. Much the fullest, most exact, most useful compilation of herbal material. Gigantic alphabetical encyclopedia, from aconite to zedoary, gives botanical information, medical properties, folklore, economic uses, much else. Indispensable to serious reader. 161 illustrations. 888pp. 6½ x 9¼. 2-vol. set. (Available in U.S. only.) Vol. I: 22798-7
Vol. II: 22799-5

HIDDEN TREASURE MAZE BOOK, Dave Phillips. Solve 34 challenging mazes accompanied by heroic tales of adventure. Evil dragons, people-eating plants, blood-thirsty giants, many more dangerous adversaries lurk at every twist and turn. 34 mazes, stories, solutions. 48pp. 8¼ x 11. 24566-7

LETTERS OF W. A. MOZART, Wolfgang A. Mozart. Remarkable letters show bawdy wit, humor, imagination, musical insights, contemporary musical world; includes some letters from Leopold Mozart. 276pp. 5⅜ x 8½. 22859-2

BASIC PRINCIPLES OF CLASSICAL BALLET, Agrippina Vaganova. Great Russian theoretician, teacher explains methods for teaching classical ballet. 118 illustrations. 175pp. 5⅜ x 8½. 22036-2

THE JUMPING FROG, Mark Twain. Revenge edition. The original story of The Celebrated Jumping Frog of Calaveras County, a hapless French translation, and Twain's hilarious "retranslation" from the French. 12 illustrations. 66pp. 5⅜ x 8½. 22686-7

BEST REMEMBERED POEMS, Martin Gardner (ed.). The 126 poems in this superb collection of 19th- and 20th-century British and American verse range from Shelley's "To a Skylark" to the impassioned "Renascence" of Edna St. Vincent Millay and to Edward Lear's whimsical "The Owl and the Pussycat." 224pp. 5⅜ x 8½. 27165-X

COMPLETE SONNETS, William Shakespeare. Over 150 exquisite poems deal with love, friendship, the tyranny of time, beauty's evanescence, death and other themes in language of remarkable power, precision and beauty. Glossary of archaic terms. 80pp. 5³⁄₁₆ x 8¼. 26686-9

THE BATTLES THAT CHANGED HISTORY, Fletcher Pratt. Eminent historian profiles 16 crucial conflicts, ancient to modern, that changed the course of civilization. 352pp. 5⅜ x 8½. 41129-X

THE WIT AND HUMOR OF OSCAR WILDE, Alvin Redman (ed.). More than 1,000 ripostes, paradoxes, wisecracks: Work is the curse of the drinking classes; I can resist everything except temptation; etc. 258pp. 5⅜ x 8½. 20602-5

SHAKESPEARE LEXICON AND QUOTATION DICTIONARY, Alexander Schmidt. Full definitions, locations, shades of meaning in every word in plays and poems. More than 50,000 exact quotations. 1,485pp. 6½ x 9¼. 2-vol. set.
Vol. 1: 22726-X
Vol. 2: 22727-8

SELECTED POEMS, Emily Dickinson. Over 100 best-known, best-loved poems by one of America's foremost poets, reprinted from authoritative early editions. No comparable edition at this price. Index of first lines. 64pp. 5$\frac{3}{16}$ x 8¼. 26466-1

THE INSIDIOUS DR. FU-MANCHU, Sax Rohmer. The first of the popular mystery series introduces a pair of English detectives to their archnemesis, the diabolical Dr. Fu-Manchu. Flavorful atmosphere, fast-paced action, and colorful characters enliven this classic of the genre. 208pp. 5$\frac{3}{16}$ x 8¼. 29898-1

THE MALLEUS MALEFICARUM OF KRAMER AND SPRENGER, translated by Montague Summers. Full text of most important witchhunter's "bible," used by both Catholics and Protestants. 278pp. 6⅝ x 10. 22802-9

SPANISH STORIES/CUENTOS ESPAÑOLES: A Dual-Language Book, Angel Flores (ed.). Unique format offers 13 great stories in Spanish by Cervantes, Borges, others. Faithful English translations on facing pages. 352pp. 5⅜ x 8½. 25399-6

GARDEN CITY, LONG ISLAND, IN EARLY PHOTOGRAPHS, 1869–1919, Mildred H. Smith. Handsome treasury of 118 vintage pictures, accompanied by carefully researched captions, document the Garden City Hotel fire (1899), the Vanderbilt Cup Race (1908), the first airmail flight departing from the Nassau Boulevard Aerodrome (1911), and much more. 96pp. 8⅞ x 11¾. 40669-5

OLD QUEENS, N.Y., IN EARLY PHOTOGRAPHS, Vincent F. Seyfried and William Asadorian. Over 160 rare photographs of Maspeth, Jamaica, Jackson Heights, and other areas. Vintage views of DeWitt Clinton mansion, 1939 World's Fair and more. Captions. 192pp. 8⅞ x 11. 26358-4

CAPTURED BY THE INDIANS: 15 Firsthand Accounts, 1750-1870, Frederick Drimmer. Astounding true historical accounts of grisly torture, bloody conflicts, relentless pursuits, miraculous escapes and more, by people who lived to tell the tale. 384pp. 5⅜ x 8½. 24901-8

THE WORLD'S GREAT SPEECHES (Fourth Enlarged Edition), Lewis Copeland, Lawrence W. Lamm, and Stephen J. McKenna. Nearly 300 speeches provide public speakers with a wealth of updated quotes and inspiration–from Pericles' funeral oration and William Jennings Bryan's "Cross of Gold Speech" to Malcolm X's powerful words on the Black Revolution and Earl of Spenser's tribute to his sister, Diana, Princess of Wales. 944pp. 5⅜ x 8⅜. 40903-1

THE BOOK OF THE SWORD, Sir Richard F. Burton. Great Victorian scholar/adventurer's eloquent, erudite history of the "queen of weapons"–from prehistory to early Roman Empire. Evolution and development of early swords, variations (sabre, broadsword, cutlass, scimitar, etc.), much more. 336pp. 6⅛ x 9¼.
25434-8

AUTOBIOGRAPHY: The Story of My Experiments with Truth, Mohandas K. Gandhi. Boyhood, legal studies, purification, the growth of the Satyagraha (nonviolent protest) movement. Critical, inspiring work of the man responsible for the freedom of India. 480pp. 5⅜ x 8½. (Available in U.S. only.) 24593-4

CELTIC MYTHS AND LEGENDS, T. W. Rolleston. Masterful retelling of Irish and Welsh stories and tales. Cuchulain, King Arthur, Deirdre, the Grail, many more. First paperback edition. 58 full-page illustrations. 512pp. 5⅜ x 8½. 26507-2

THE PRINCIPLES OF PSYCHOLOGY, William James. Famous long course complete, unabridged. Stream of thought, time perception, memory, experimental methods; great work decades ahead of its time. 94 figures. 1,391pp. 5⅜ x 8½. 2-vol. set.
Vol. I: 20381-6 Vol. II: 20382-4

THE WORLD AS WILL AND REPRESENTATION, Arthur Schopenhauer. Definitive English translation of Schopenhauer's life work, correcting more than 1,000 errors, omissions in earlier translations. Translated by E. F. J. Payne. Total of 1,269pp. 5⅜ x 8½. 2-vol. set. Vol. 1: 21761-2 Vol. 2: 21762-0

MAGIC AND MYSTERY IN TIBET, Madame Alexandra David-Neel. Experiences among lamas, magicians, sages, sorcerers, Bonpa wizards. A true psychic discovery. 32 illustrations. 321pp. 5⅜ x 8½. (Available in U.S. only.) 22682-4

THE EGYPTIAN BOOK OF THE DEAD, E. A. Wallis Budge. Complete reproduction of Ani's papyrus, finest ever found. Full hieroglyphic text, interlinear transliteration, word-for-word translation, smooth translation. 533pp. 6½ x 9¼. 21866-X

MATHEMATICS FOR THE NONMATHEMATICIAN, Morris Kline. Detailed, college-level treatment of mathematics in cultural and historical context, with numerous exercises. Recommended Reading Lists. Tables. Numerous figures. 641pp. 5⅜ x 8½. 24823-2

PROBABILISTIC METHODS IN THE THEORY OF STRUCTURES, Isaac Elishakoff. Well-written introduction covers the elements of the theory of probability from two or more random variables, the reliability of such multivariable structures, the theory of random function, Monte Carlo methods of treating problems incapable of exact solution, and more. Examples. 502pp. 5⅜ x 8½. 40691-1

THE RIME OF THE ANCIENT MARINER, Gustave Doré, S. T. Coleridge. Doré's finest work; 34 plates capture moods, subtleties of poem. Flawless full-size reproductions printed on facing pages with authoritative text of poem. "Beautiful. Simply beautiful."–*Publisher's Weekly.* 77pp. 9¼ x 12. 22305-1

NORTH AMERICAN INDIAN DESIGNS FOR ARTISTS AND CRAFTSPEOPLE, Eva Wilson. Over 360 authentic copyright-free designs adapted from Navajo blankets, Hopi pottery, Sioux buffalo hides, more. Geometrics, symbolic figures, plant and animal motifs, etc. 128pp. 8⅜ x 11. (Not for sale in the United Kingdom.) 25341-4

SCULPTURE: Principles and Practice, Louis Slobodkin. Step-by-step approach to clay, plaster, metals, stone; classical and modern. 253 drawings, photos. 255pp. 8⅛ x 11. 22960-2

THE INFLUENCE OF SEA POWER UPON HISTORY, 1660–1783, A. T. Mahan. Influential classic of naval history and tactics still used as text in war colleges. First paperback edition. 4 maps. 24 battle plans. 640pp. 5⅜ x 8½. 25509-3

THE STORY OF THE TITANIC AS TOLD BY ITS SURVIVORS, Jack Winocour (ed.). What it was really like. Panic, despair, shocking inefficiency, and a little heroism. More thrilling than any fictional account. 26 illustrations. 320pp. 5⅜ x 8½. 20610-6

FAIRY AND FOLK TALES OF THE IRISH PEASANTRY, William Butler Yeats (ed.). Treasury of 64 tales from the twilight world of Celtic myth and legend: "The Soul Cages," "The Kildare Pooka," "King O'Toole and his Goose," many more. Introduction and Notes by W. B. Yeats. 352pp. 5⅜ x 8½. 26941-8

BUDDHIST MAHAYANA TEXTS, E. B. Cowell and others (eds.). Superb, accurate translations of basic documents in Mahayana Buddhism, highly important in history of religions. The Buddha-karita of Asvaghosha, Larger Sukhavativyuha, more. 448pp. 5⅜ x 8½. 25552-2

ONE TWO THREE . . . INFINITY: Facts and Speculations of Science, George Gamow. Great physicist's fascinating, readable overview of contemporary science: number theory, relativity, fourth dimension, entropy, genes, atomic structure, much more. 128 illustrations. Index. 352pp. 5⅜ x 8½. 25664-2

EXPERIMENTATION AND MEASUREMENT, W. J. Youden. Introductory manual explains laws of measurement in simple terms and offers tips for achieving accuracy and minimizing errors. Mathematics of measurement, use of instruments, experimenting with machines. 1994 edition. Foreword. Preface. Introduction. Epilogue. Selected Readings. Glossary. Index. Tables and figures. 128pp. 5⅜ x 8½. 40451 X

DALÍ ON MODERN ART: The Cuckolds of Antiquated Modern Art, Salvador Dalí. Influential painter skewers modern art and its practitioners. Outrageous evaluations of Picasso, Cézanne, Turner, more. 15 renderings of paintings discussed. 44 calligraphic decorations by Dalí. 96pp. 5⅜ x 8½. (Available in U.S. only.) 29220-7

ANTIQUE PLAYING CARDS: A Pictorial History, Henry René D'Allemagne. Over 900 elaborate, decorative images from rare playing cards (14th–20th centuries): Bacchus, death, dancing dogs, hunting scenes, royal coats of arms, players cheating, much more. 96pp. 9¼ x 12¼. 29265-7

MAKING FURNITURE MASTERPIECES: 30 Projects with Measured Drawings, Franklin H. Gottshall. Step-by-step instructions, illustrations for constructing handsome, useful pieces, among them a Sheraton desk, Chippendale chair, Spanish desk, Queen Anne table and a William and Mary dressing mirror. 224pp. 8⅛ x 11¼. 29338-6

THE FOSSIL BOOK: A Record of Prehistoric Life, Patricia V. Rich et al. Profusely illustrated definitive guide covers everything from single-celled organisms and dinosaurs to birds and mammals and the interplay between climate and man. Over 1,500 illustrations. 760pp. 7½ x 10⅛. 29371-8

Paperbound unless otherwise indicated. Available at your book dealer, online at **www.doverpublications.com**, or by writing to Dept. GI, Dover Publications, Inc., 31 East 2nd Street, Mineola, NY 11501. For current price information or for free catalogues (please indicate field of interest), write to Dover Publications or log on to **www.doverpublications.com** and see every Dover book in print. Dover publishes more than 500 books each year on science, elementary and advanced mathematics, biology, music, art, literary history, social sciences, and other areas.